SEE AMERICA

SEE AMERICA

The Politics and Administration of
Federal Tourism Promotion, 1937–1973

MORDECAI LEE

Cover image: Arches National Park, See America campaign, USTB poster.
Frank S. Nicholson, New York City: Federal Arts Project, WPA, about 1940.

Published by State University of New York Press, Albany

© 2020 State University of New York

All rights reserved

No part of this book may be used or reproduced in any manner whatsoever without written permission. No part of this book may be stored in a retrieval system or transmitted in any form or by any means including electronic, electrostatic, magnetic tape, mechanical, photocopying, recording, or otherwise without the prior permission in writing of the publisher.

For information, contact State University of New York Press, Albany, NY
www.sunypress.edu

Library of Congress Cataloging-in-Publication Data

Names: Lee, Mordecai, 1948– author.
Title: See America : The politics and administration of federal tourism promotion, 1937–1973 / Mordecai Lee.
Description: Albany : State University of New York Press, 2020. | Includes bibliographical references and index.
Identifiers: LCCN 2019036157 | ISBN 9781438478098 (hardcover : alk. paper) | ISBN 9781438478081 (pbk. : alk. paper) | ISBN 9781438478104 (ebook)
Subjects: LCSH: Tourism—United States—History. | Tourism—Government policy—United States. | United States. Department of the Interior—History. | United States. Travel bureau. [from old catalog]
Classification: LCC G155.U6 L43 2020 | DDC 917.304/92—dc23
LC record available at https://lccn.loc.gov/2019036157

10 9 8 7 6 5 4 3 2 1

For my sisters, who have kept me on the straight and narrow (in birth order, not playing favorites!): Riva Nolley, Tahirih Lee-Sursin, and Shirin Coleman

Contents

Preface		ix
List of Abbreviations		xv
Introduction		1
Chapter 1	The Tourism Industry and Big Government: Origins of the US Travel Bureau, 1930–1936	15
Chapter 2	The US Tourist Bureau: Birth by Administrative Action, 1937	21
Chapter 3	The US Travel Bureau: Renamed and Expanded, 1938	27
Chapter 4	When Business Liked (Part of) the New Deal, 1939	35
Chapter 5	Congress Decides It Sometimes Likes Agency PR: Statutory Creation, 1940	49

color gallery follows page 62

Chapter 6	Promoting Tourism during a National Emergency, 1941	63
Chapter 7	Travel Promotion in Wartime? 1942–1943	81
Chapter 8	Postwar Revival: Interior's US Travel Division, 1946–1949	101

Chapter 9 Last Try: Interior's Office of Travel, 1968–1973	113
Conclusion	135
Notes	155
Bibliography	199
Index	215

Preface

In 2017, the US Postal Service issued a set of stamps commemorating Depression-era federal posters designed by the graphic artists of the Works Progress Administration (WPA). Selecting ten of the best and most iconic WPA posters, one was for an obscure federal agency called the United States Travel Bureau (USTB). WPA is one of the relatively well-known legacies of Franklin Roosevelt's New Deal, in part, due to its role in art. It hired artists, writers, playwrights, filmmakers, and actors who were out of work due to the Great Depression. Why? As Harry Hopkins, WPA's director, famously said, "Artists have to eat, too." His quip also reflected FDR's strong political preference that it was much better to pay unemployed Americans during the Great Depression to *work* rather than to be on the dole. The key was to move away from the politically charged subject of relief, of supposedly merely handing out money to people who did nothing in return. That kind of welfare only reinforced the conservative political stereotyping of lazy and undeserving recipients.

The way FDR and Hopkins saw it, there was little difference between art and other categories of work that WPA paid people to do. After all, artists created lasting products just as much as the more visible and concrete WPA projects, such as post offices, forests, parks, dams, courthouses, soil conservation, and highways. WPA's artists created murals in public buildings, books (including a series of state guides), handicrafts, paintings, plays, and sculptures. In particular, WPA sought artistic products that contradicted the image of highbrow and elitist art. It wanted art for the people. This is one of the reasons why its arts program included poster design.

Among other projects, WPA graphic artists designed posters in response to requests from federal and other public agencies. This 2017 stamp was from a poster that was part of a public relations campaign of the USTB to promote increased domestic tourism. The name of the campaign was "See America." The poster showed a rider on horseback admiring the outdoor scenery of forests and snow-capped mountains of Montana. Hence, the text on the poster (and stamp) was "See America / Welcome to Montana / United States Travel Bureau." (See figure 1.) The Montana poster was one of a series of about seven that WPA graphic artists designed for USTB's See America campaign. Some of the others promoted famous landscapes in national parks, including Arches National Park in Utah and the Carlsbad Caverns in New Mexico. While many Americans continue to be aware of WPA buildings that are still in use, as well as having a vague awareness of its arts contributions, few would link WPA to the largely forgotten USTB.

USTB not only tried to stimulate domestic tourism through public relations (PR) projects like posters and its See America campaign, but it also played a quiet role in racial matters. The bureau used targeted PR to encourage African Americans to travel and to enjoy seeing the country as much as whites. This effort was admirable and important, particularly given that FDR sought to avoid taking actions to inflame the white power structure. As president, Roosevelt was very cautious on promoting civil rights. While African American voters supported him lopsidedly, their votes were counterbalanced by the outsize role of southern states in the Electoral College, the inordinate influence of southerners in Congress due to the seniority system, and the filibuster rule in the US Senate. These politicians would not only oppose any civil rights initiatives of FDR's that required legislation or appropriations, but they could counterstrike by stifling the president's other unrelated legislative proposals. That explains in part why FDR's relatively modest and reactive measures to support civil rights were almost exclusively limited to executive orders and other internal management directives. He could issue them without a role for Congress, let alone its approval. One of the long-running political conflicts during WWII was his executive order creating the Fair Employment Practices Committee (FEPC) and the persistent congressional efforts to abolish or at least defund it (Lee 2018, 95–99; 2016, 99–101). In the context of FDR's deliberate low profile and passivity on equal rights, USTB stands out. It proactively sought to reach an audience of African Americans as a segment of the general population for PR programs to encourage domestic tourism and travel.

For example, in 1936–37, an Interior Department study of parks and recreation facilities nationally included an examination of the "needs of Negro" citizens (US National Park Service [NPS] 1938, 7). In 1938, USTB's New

York office hired Charles McDowell, who had been the first African American agent for the Greyhound bus line. The bureau appointed him as a full-time employee (along with two African American women as assistants) to promote travel and tourism by African Americans. One of his assignments was to research "the problems attendant upon Negro travel throughout the United States."[1] The next year, based on information McDowell collected, USTB published a twelve-page *Directory of Negro Hotels and Guest Houses in the United States* (USTB 1939e). Throughout 1939–40, he routinely contributed a column titled "Travel Notes" in USTB's twice-monthly newsletter that was widely distributed throughout the eastern United States. In his byline, he was identified as head of USTB's Division of Negro Activities. USTB's *Official Bulletin*, which had national circulation, contained an article about its role in the publication of the new edition of the *Negro Motorist Green Book* (Green 1940). The title page of the book stated that it was "Prepared in cooperation with The United States Travel Bureau." MacDowell also wrote an article in it about traveling in the South (McDowell 1940). In 1941, USTB released a seventeen-page updated and expanded edition of its *Directory of Negro Hotels and Guest Houses* (USTB 1941c). These openly conducted administration activities in support of African Americans were nonetheless low key enough that they never attracted the attention of racists in and out of Congress.

There was another oddity related to USTB. Historical narratives often focus on the persistent and insistent hostility that business had to the New Deal and to FDR. Roosevelt was depicted as the enemy of free enterprise and capitalism. His reforms, such as increasing regulation of Wall Street and expansion of the bureaucracy to protect workers' rights, were greeted with stark hostility from chambers of commerce and business organizations. Those with economic advantages referred to FDR as a "traitor to his class," indicating their implacable hatred of him (Brands 2008). Yet, oddly, business was an enthusiastic supporter of USTB, even though it was part of FDR's presidency and a new federal program, expanding into an arena where it had never ventured before. The travel sector of the economy was substantial (perhaps a quarter of the economy), including resorts, hotels, travel agents, gas stations, and transportation companies, such as railroads, bus companies, and ships. By the late 1930s, the transportation subsector expanded to a new and growing industry: commercial airlines. All these businesses openly supported USTB and frequently lobbied Congress to pass legislation creating it and, once Congress did so, to continue funding it generously. As with civil rights, USTB's support from business was unusual for the times.

I came to become interested in the forgotten history of USTB through a circuitous route. As a senior at the University of Wisconsin-Madison in 1969–70, my advisor was political science professor James McCamy. When I was struggling with what to do after graduation, he suggested I consider graduate work in public administration. I didn't even know that there was such a profession, let alone an academic field of study. "Why not start graduate studies working toward an MPA degree," he asked, "and see if you like it?" As I warmed to the idea, he told me that his first research interest as an academic was the study of government public relations. During my subsequent doctoral studies in public administration at Syracuse University and later academic career at the University of Wisconsin-Milwaukee, I explored multiple aspects of government PR in books, articles, chapters, and encyclopedia entries.

About five years ago, while researching a different government PR program, I happened upon a reference to the USTB. It was an entity in the Department of Interior during the late 1930s and early 1940s. That was all news to me; I had never heard of it before. However, it seemed quite intriguing. Here was a government agency whose purpose was to promote domestic tourism through various public relations efforts. Its mission made sense on several levels. First, to recover from the Great Depression, increased travel spending was one way to lift the economy. Second, Roosevelt had been a strong supporter of conservation and felt that increased visits to national parks and other federal recreational venues could contribute to public support for those spending priorities—a nice antidote to the near-constant criticism by the right of New Deal spending and "big government." Third, domestic tourism could promote patriotism and love of country, something increasingly important given the war clouds on the horizon and FDR's pivot to strengthening national defense, particularly after the German invasion of Poland in September 1939.

What struck me the most was that USTB seemed to be something of an exception to the rule of government PR. Congress, especially the conservative coalition, consistently criticized public relations by federal agencies. Federal agencies should be seen and not heard, they claimed. Civil servants were to be in the business of quietly and efficiently implementing the programs that Congress approved and funded—nothing more. In particular, public administration should not be promoting itself by spending money on PR or seeking public support. Yet here was a federal agency, I gradually learned, that was created by a law approved by Congress and funded through its annual appropriation process. And it was solely in the

PR business. It did not *do* anything other than PR. I just had to find out more about this oddity. This book is the result of trying to learn about this relatively unknown aspect of federal public relations.

As is always the case for researchers, I could not have succeeded in reconstructing the story of USTB without archivists, librarians, and new online sites and search engines. Archivists at the National Archives II and its Center for Legislative Archives, the Manuscript Division of the Library of Congress, as well as the FDR, Kennedy, and LBJ Presidential Libraries were all extraordinarily helpful and patient. The same was true for staffers at archival collections held by Clemson University and the University of Alaska-Anchorage. Also helpful were librarians at the Department of Interior's Library and the Government Documents Department of the Wisconsin Historical Society. I regret that I cannot list each by name because the list would be too long. Instead, I salute these wonderful professionals who helped me. Back home, the Inter-Library Loan Department of my university's Golda Meir Library was an extraordinary source of assistance. Beth Kucera, its head, performed what can only be considered miracles in obtaining materials I asked for. She had a never-say-die attitude, even for published materials held by other libraries (sometimes only one library) that were ostensibly noncirculating. I am deeply appreciative of her ingenuity and persistence. Of all the information professionals who helped me, she surely contributed the most to this project. Also, my thanks for help locating a USTB map in the collection of the American Geographical Society at my library.

Looking back on the advances in online information technology that occurred during my academic career, I am in awe of how much researchers have benefitted from online databases and search engines. Sites such as JSTOR, Google Scholar, WorldCat/OCLC, ProQuest Historical Newspapers, and NewspaperArchive.com permit searches for specific names and subjects with a detail and comprehensiveness that was unimaginable when I was in graduate school in the 1970s. I consider myself lucky that my academic career overlapped with these wonderful new tools. (Not to mention the miracle of word processing—once you fixed a mistake, it *stayed* fixed. That was not the case for the multiple typed and retyped versions of my dissertation in 1975.)

It is a pleasure to thank SUNY Press for publishing this book, my fourth with it over a fifteen-year period. I am deeply appreciative to all of the press staff who have worked with me. In particular, I must single out senior acquisitions editor Michael Rinella, who shepherded all four books to fruition. I consider myself lucky to have been under his tutelage. He roots for his authors and wants the best for them. I am thankful to have benefited from his welcoming attitude, patience, and reassuring touch.

Finally, a note on the referencing style used in the book. Generally, the parenthetical citation style of author-year-page is a very concise way to cite published sources. However, it is quite prolix when referencing several other categories of sources. For example, the parenthetical style is very cumbersome for citing archival sources. For those sources, the endnote style of referencing facilitates brevity. That is also the case for journalistic sources (especially when the article is not bylined), and other nonacademic sources. Therefore, citations for the latter categories are in endnotes. I have also used endnotes for tangential discussions and asides. This dual referencing style permits the most condensed and compact approach to citing sources. While unorthodox, it is economical and avoids excess verbiage that would occur if dogmatically being limited to only one referencing style. SUNY Press had permitted me to use this approach in two previous books, and I appreciate its willingness to accede to it again for this volume.

As always, notwithstanding the help and assistance of so many people, any remaining errors in this book are mine alone.

Abbreviations

AP   Associated Press (news wire service)

BOB   Bureau of the Budget (an agency in the Executive Office of the President)

CLA   Center for Legislative Archives (maintained for Congress by the National Archives, an executive-branch agency)

CR   *Congressional Record*

CSM   *Christian Science Monitor* (Boston-based afternoon newspaper)

CT   *Chicago Tribune*

FDR   Franklin Delano Roosevelt

FY   Fiscal Year. During the period discussed in this book, the federal fiscal year started on July 1 and ended on June 30 of the next calendar year. Fiscal years were named by the year they ended in. For example, FY 1941 started on July 1, 1940 and ended on June 30, 1941. Normally a president submitted his budget plan to Congress in January, giving the legislative branch six months to pass appropriations bills. So the period from January to June was the time when agencies faced budgeting scrutiny on Capitol Hill (and, before that, by BOB). *Note:* In 1974, fiscal years were bumped forward by a quarter. In this revised budgeting calendar, the fiscal year started on October 1 and ended on September 30. The purpose was to give Congress an additional three months to pass funding legislation after receiving the president's plan early in the calendar year. (It didn't help.)

GPO	Government Printing Office
HC	*Hartford [CT] Courant*
LAT	*Los Angeles Times*
NA II	National Archives, site II, College Park (MD)
NPS	National Park Service, a bureau of the Department of Interior
NYHT	*New York Herald Tribune*
NYT	*New York Times*
OB	*Official Bulletin*, published by the United States Travel Bureau
PR	public relations (usually treated as a compound noun or adjective and as singular)
RG	Record Group (filing category used by the National Archives)
Stat.	*United States Statutes at Large*
USTB	United States Travel Bureau (briefly in 1937–38: US *Tourist Bureau*)
USTD	United States Travel Division
WP	*Washington Post*
WS	*Washington Star* (afternoon newspaper, except Sunday)
WWII	World War II

Introduction

Overview of the US Travel Bureau

From its earliest operation in 1937 until 1942, USTB was a novel and nearly unique agency in the American federal government. Until its creation, the common view was that tourism promotion was not a proper activity for the federal executive branch. It belonged to private businesses that benefited from it and to state and local tourism offices for those regions that economically were significantly dependent on visitors. For example, urban chambers of commerce had organized and promoted city festivals to attract tourists (Cocks 2001) and railroads advertised the scenic locations passengers could visit (Shaffer 2001, chap. 2).

USTB broke that template. It was the first formal effort in the federal government to promote domestic tourism. Its task was to do something Washington had not done before or considered to be within its ken: promoting travel and tourism to the citizenry-at-large. The bureau did this with a full-throated and robust public relations effort, including publications, posters, campaigns, radio programs, advertising, conferences, and newsletters. These PR programs co-existed with Congress's longstanding and general opposition to executive-branch PR, which had been heightened in reaction to FDR's significant expansion of the public relations programs of federal agencies (Lee 2011).

The bureau first came into existence in 1937, when Interior Secretary Harold Ickes established it through a secretarial order. Its initial name was the US Tourist Bureau. Jockeying with the Commerce Department for primacy in promoting tourism, Ickes eventually persuaded Secretary of Commerce Harry Hopkins to yield in the turf battle and agree that domestic travel promotion should go to Interior. Then, with the enthusiastic support of the business sector, the Seventy-sixth Congress (1939–40) passed legislation authorizing a more formal status for USTB. This law also made the bureau eligible for funding in the annual departmental appropriation bill. FDR signed the bill into law in mid-July 1940, though the bill signing was understandably drowned out by the political news of his nomination for a third term by the Democratic Party's convention in Chicago and by the war news from Europe.

After Pearl Harbor, the agency had an on-and-off existence, still operating nominally until 1943, reviving in 1946, dying again in 1949, reviving again in 1968, losing its funding in 1971, and finally losing its statutory authorization in 1973.

Literature Review: American Public Administration History

Contemporary literature in US public administration is heavily tilted to current-day practices, case studies, empirical research, behavioral research, and quantitative studies. These contribute to the scholarly body of knowledge, yet they are often wholly or largely disconnected from historical events. This often leads to an inaccurate sense of modern practice occurring in a vacuum, even of newness, which can be quite incorrect. Therefore, historical knowledge can help contemporary researchers by adding perspective and context to the subjects being examined.

The study of American public administration history has been a relatively modest genre within the discipline, but it has gradually been growing in scope. Published historical research since 2000 tends to focus mostly on horizontal subjects with broad and overall perspectives, including the Founding Fathers (Bertelli and Lynn 2006; Newbold 2010), gender (Gabriele 2015; Stivers 2000), pedagogy (Raadschelders 2017), Progressivism (Durant 2014; McDonald 2010), performance measurement (Williams 2002), role of Congress (Rosenbloom 2000), public law (Mashaw 2012), civil service (Pfiffner and Brook 2000; White 2003), budgeting (Meyers and Rubin 2011), management fads (Schachter 2017), American political development (Hoffer 2007), and even counterfactual history (Lee 2005a). President Franklin Roosevelt's seminal role in public administration also continues to be of central interest to historians, with recent examples including the Brownlow Committee report, its subsequent implementation, enhancement of personnel management, and FDR's first budget director (Lee 2016; 2018; Newbold and Rosenbloom 2007; Newbold and Terry 2006; Zelizer 2012, chap. 7).

However, there is less historical research using a vertical silo approach, examining individual line agencies in the executive branch. This approach was very popular early in the twentieth century, when the Washington-based Institute for Government Research (a predecessor to the Brookings Institution) began publishing a book series of sixty-six agency profiles called Service Monographs of the United States Government. The first installment was on the Interior Department's Geological Survey in 1918, and the last volume was published in 1934 on the Veterans Administration (*US Geological Survey* 1918; Weber and Schmeckebier 1934). Later, commercial publishing house Praeger released a series of about forty-five volumes on federal departments and agencies from 1967 to 1976, but these were oriented to a more popular readership. There has been something of a modest revival in this somewhat old-fashioned and less glamorous approach to public administration history, although not framed as such. They include examinations of the history of the Bureau of Indian Affairs (Rockwell 2010), the General Land Office (Stivers 2011), NASA (Lambright 2017; 2007), the Food and Drug Administration (Carpenter 2010), and a bureau in the Federal Trade Commission (Paulter 2015).

There is also a growing literature on the history of tourism and travel promotion, both in the United States and abroad. These include Apostle (2001), Baranowski and Furlough (2001), Dawson (2004), Dubinsky (1999), Furlough (1998), Semmens (2005), Shaffer (2001), and Wrobel and Long (2001).

Prospective Significance of the US Travel Bureau

The scholarly literature of American public administration has not included an overall or in-depth examination of USTB. In fact, it is almost entirely invisible. However, the bureau has been of passing interest in the literature of other academic disciplines, particularly the emerging literature on tourism and parks (Berger 2006; Berkowitz 2001; Dawson 2011; Duchemin 2009; Popp 2012; Swain 1972). Other academic research has discussed (or at least glancingly alluded to) USTB vis-à-vis African Americans (Armstead 2005; Sorin 2009), congressional committee jurisdictions (King 1994), WPA writers and artists (Griswold 2016; Pillen 2008), and a biography of Hollywood cowboy Gene Autry (Duchemin 2016). Also, a dissertation in history examined the cultural politics of representations of the American national identity from 1930 to 1960 and included a well-researched chapter on USTB (McLennan 2015, chap. 1).

This inquiry seeks to add to the historical literature in two ways. First, it is a vertical study of one federal agency, representative of the larger transformation of the federal administrative apparatus during FDR's presidency.

The USTB, founded in 1937, can be seen as an example of FDR's expansion in the scope of the federal government. Up to that point, tourism and travel promotion had never been a responsibility or departmental mission in Washington. FDR's administration pushed hard to expand the functions of the executive branch to include a tourism promotion office. In that respect, USTB is an exemplar for the broadening of the scope of the federal government during the New Deal in the 1930s and 40s.

Yet, beyond this generic rationale for a public administration case study that had previously been largely overlooked, USTB is of interest for several other distinct reasons. It stands out because, unlike so much of FDR's expansion of the federal executive branch, its legal status was (eventually) authorized by an act of Congress. This statute was in stark contrast to the many alphabet agencies he created with their legality based solely on executive orders (Lee 2018, 24; 2016, 100). How and why was USTB an exception? The answer leads to another unusual aspect of its existence: unlike the strong animosity and even hatred of FDR and the New Deal by American business and its allies on Capitol Hill, USTB was a program that business liked, wanted, and lobbied for. These strange political bedfellows stood out in an era of the private sector's seemingly implacable opposition to anything relating to FDR and the New Deal.

Travel Promotion as Government PR

Another unusual element of USTB is that the bureau did not produce any kind of tangible product. Unlike most of public administration at the time, it did not create anything, such as highways, parks, maps, and ships; did not run anything, such as schools, Indian reservations, dams, and grazing lands; did not distribute money, such as relief, farm subsidies, and construction grants to local governments; and did not produce scientific research, such as on plants and insects, labor statistics, and drug safety. All USTB did was disseminate information. It was simply and wholly a PR agency.

There were a few limited precedents to such an unusual activity in American public administration, such as USDA distributing information to farmers to stabilize and increase their incomes or the Department of Commerce notifying businesses of export opportunities. In those more traditional cases, the circulation of information was largely limited to specific slices of the economy or to well-defined and attentive audiences, such as farmers and businesses. The information was not of interest to the public at large, nor did it benefit the citizenry in any direct or useful way. USTB was different. Its informational products and outputs were geared to the undifferentiated public, to any resident (or even foreigner) who might be a potential traveler in the United States.

In some of those comparable examples of agency PR, information that the federal agency would circulate and publicize was almost always prepared and generated in-house by the agency itself. For example, the weather bureau researched meteorological patterns and then, based on its data gathering, released forecasts that interested the public at large. Some of the information USTB disseminated was generated in-house, such as directories, monthly bulletins, maps, calendars of events, and travel posters. However, more often than not, the information it possessed and distributed was provided by entities outside the bureau, even outside the federal government. These materials came from for-profit corporations, such as railroads, gasoline companies, shipping lines, and airlines; from nonprofit chambers of commerce and other business associations; and from the tourist promotion offices of states and regions with a particular interest in attracting vacationers. In those cases, the USTB received materials from those outside sources and then engaged in PR to circulate them. That made the bureau something of an informational way station, a go-between service that added the imprimatur of the federal government on materials generated by third parties. This was very unusual role for the federal civil service.

Finally, USTB's information outputs of the agency were not merely factual, on the level of, say, crop reports and employment statistics. Instead, they were intended to be persuasive. How about taking a vacation within the United States? How about visiting these interesting places? These PR outputs were aimed at influencing the *behavior* of citizens, of convincing them to travel in the United States. This could just as easily be called propaganda because its intent was to affect economic and leisure activities of the public at large. This was contradistinction to the more precedent-based informational dissemination activities of the traditional executive branch. For example, the NWS distributes free and neutral information but with no explicit persuasive bent. FYI, it might rain tomorrow, but it is up to you to decide if you want to take an umbrella when you leave for work or school. In all, USTB's *raison d'être* was pure public relations. PR was its sole mission. Persuasion and modified behavior were the intended outcomes. Success would accrue to the economic benefit of others, not the federal government nor federal taxpayers directly.

As a general rule, Congress, conservatives, and the news media were strongly antipathetic to federal PR, denouncing it as self-serving propaganda, back-patting, and insidious indirect lobbying of Congress by building a positive image with voters. With effective PR, an agency could increase its autonomy from political oversight and control (Carpenter 2001; Lee 2011). Oddly, USTB faced few to none of those standard political accusations. If anything, the usual sources of criticism of agency PR were pressing the bureau to do more PR, more promotion, more information dissemination. The sky was the limit, it seemed. USTB was the exception to the political rule that federal agencies should be seen but not heard.

Published Errors about the Travel Bureau

For the relatively modest extant literature on USTB, there are an above-average number of errors and misstatements about its life and death that deserve to be corrected, so that future researchers will have a more accurate account. These errors start with the agency's beginnings and continue through to its demise.

Historical summaries of federal tourism policy and administration sometimes wholly omit USTB as the first federal tourism promotion effort. Edgell reported the first "formal legislated involvement" by the federal government in encouraging foreigners to visit the United States was in 1961 (1992, 595). This ignored USTB's early international work, particularly the formal congressional appropriation in 1940 for the bureau to prepare brochures in Spanish and Portuguese for distribution in Latin America. Airey's review of federal tourism administration also wholly overlooked USTB, starting his historical review with the 1961 law signed by President Kennedy (1984, 273). Similarly, a 2019 textbook on tourism dismissed everything that happened in the federal government before the 1970s with the observation that "tourism in the 1950s and 1960s received very little policy attention from the U.S. government" (Edgell and Swanson 2019, 191). USTB's existence in the 1930s and '40s did not merit even a passing mention.

Regarding its beginnings, Sellars referred to "the U.S. Travel *Division* . . . created in early 1937" (1993, 46, emphasis added). No. From 1937 to 1938, it was called the US Tourist Bureau. Beginning in 1939, its name changed to US Travel Bureau. Only when revived after WWII was it called the US Travel Division. He also misstated that the agency's purpose was "to stimulate travel to the national parks." Its mission was to stimulate all domestic travel and tourism, not just to national parks, even though it was a unit within the National Park Service (NPS). Mak accurately stated that before the 1940 law giving USTB statutory status, Interior Secretary

Harold Ickes had transferred it from NPS to the Office of the Secretary in March 1939 (2015, 27n88). However, he neglected to mention that only a year later (and before the congressional enactment), Ickes reversed himself and transferred it back to NPS.

Three publications stated that the 1940 law giving USTB statutory status was formally called the "Domestic Travel Act" (Gunn 1983, 33; Brewton and Withiam 1998, 54) or "the US Domestic Travel Act" (Egdall and Swanson 2013, 41). Actually, the law had no title, as was the common practice at the time.[2] Swain wrote that the 1940 law established USTB as "an independent" agency and, as a result, "the Park Service bowed out of the business" (1972, 318). No. The 1940 law explicitly created USTB as an entity *within* NPS. It stayed that way throughout its subsequent existence until the law was repealed in 1973. Furthermore, NPS never voluntarily "bowed out." Rather, it fought tooth and nail to retain this statutory mission and only gave up the fight after the secretary of interior decided in 1971 to cede the jurisdiction for this activity to the Commerce Department.

Berger referred to "the State Department's United States Travel Bureau" (2006, 85). Wrong. It was in the Interior Department. It received some modest funding from Congress in 1940–41 associated with a larger appropriation to the State Department to support tourism to and from Latin America. Egdall and Swanson wrote that USTB was "superseded in 1941" with the beginning of the war and "ceased" to operate (2013, 41). Similarly, Berkowitz stated that USTB "shut down after the first months of World War II" (2001, 203). Both references indicated that USTB closed shortly after the December 1941 Pearl Harbor attack and the US declaration of war. No, the bureau's planned shutdown was publicly announced a year later, in December 1942, and its (reduced) staff remained on the payroll until February 1943.

Based on the 1940 law that was still on the books, NPS tried to revive the travel bureau beginning in 1968. According to Gunn, it was "given greater funding in 1970–72. Then, in 1973 the program was again inactive" (1983, 33). No, Congress first formally (resumed) appropriated funding to it for fiscal year (FY) 1971 (July 1970–June 1971). The Nixon administration did not ask for funding for FY 1972, and Congress did not overrule that budget recommendation. That meant the travel office was defunded in mid-1971, at the end of FY 1971. More importantly, what happened to it in 1973 was not becoming "inactive." Rather, a new law that year repealed NPS's 1940 legal mission of promoting domestic travel and reassigned this role to the Commerce Department.

Due to this disproportionate number of errors in the literature, this inquiry seeks to reconstruct the history of USTB based as much as possible on primary and original sources and to minimize reliance on secondary sources that might be inaccurate.

Disciplinary Foci

This study of the USTB will likely be of interest to academic audiences in political science, public administration, American history, public relations, African American studies, tourism and leisure studies, and the environment.

In the social sciences, the story of the USTB embodies several different threads of attention and prisms of analysis, including Congress, FDR and the New Deal, bureaucratic politics, the executive branch, administrative history, governmental management and organization, political history, and policy history. As an inquiry about a relatively unknown federal bureau from birth to death, it seeks to present a full biography of USTB in the context of public administration and the federal government. Some of this agency's prominent features included an unusual and almost exclusive focus on public relations, the role of external business stakeholders particularly by pressuring Congress to support creating and funding it, and ongoing dependence on congressional decision making regarding the proper role of the federal government in tourism promotion.

A USTB biography is also about the power of a law, from germination to repeal. Once passed by Congress, the 1940 law assigning the domestic travel portfolio to the Department of the Interior was a kind of administrative and political Rock of Gibraltar. It set in place a permanent statutory assignment to the NPS in the Interior Department, along with the presumption of the continued operations of the agency. The 1940 law gave Interior a monopoly over domestic travel promotion. No other federal agency could legally engage in such activities. NPS was able to wield the law as a powerful club to maintain its possession of this mission. Furthermore, laws are not changed easily. Only if, or when, Congress would repeal or amend the law could the status quo be changed. In that sense, this is the biography of a law as well as of a federal agency.

The 1940 law and its aftermath also reflected internal centers of power on Capitol Hill. Congress had a required two-step process for all federal spending. In both houses, a bill *authorizing* expenditures for a particular purpose could only be handled by the pair of standing committees with substantive jurisdiction over the subject. For the policy area of domestic tourism promotion and its assignment to the Interior Department, there was conflict between the commerce committees that claimed this was a matter of commerce, and the interior committees because the locale of the activity was in the Interior Department. Generally, the weight of referral precedent gave travel and tourism bills to the commerce committees (King 1994, 57–58). Traditionally, authorization bills set a cap on the maximum amount that could be appropriated for the activity. As a matter of practice, Congress preferred not to authorize an unlimited amount.

The role of the standing committees in passing authorizing legislation was a counterweight to the power of the purse held by the two appropriations committees, with the House committee the more important of the two because the Constitution required that all funding bills had to start there. As a result, appropriations bills were limited to funding an activity that had already been authorized by laws controlled by the standing committees. However, such authorizations did not compel Congress to appropriate any funding for that particular authorized activity. Appropriations Committees could recommend spending less than the full authorized amount. They could not even fund the program at all. A zeroing out of funding was a de facto repeal of the authorized activity. Conversely, appropriations could not be made to any federal activity that was not already authorized by a preceding and separate law. Similarly, an appropriation bill was prohibited from amending authorizing legislation. Each funding bill was limited to authorizing decisions about the annual amount of money dedicated to that activity or agency.[3]

USTB's story is about politics writ large, including that of private-sector interests lobbying Congress, an agency seeking to promote its standing with stakeholders, cabinet secretaries competing for primacy within a presidential administration, and whether differences in party control of Capitol Hill and the White House were significant in setting USTB's role. The USTB also embodies a never-ending bureaucratic turf war between the Departments of Interior and Commerce that seemed to be the public administration equivalent of the Thirty Years War. The multiple changes in presidents and secretaries appeared to have little effect on their respective and continuing efforts for primacy.

This political, legislative, and bureaucratic wrangling also presents a case study of public policy: who makes it, how it is made, how it is implemented, and how it is changed. Just as USTB is a biography of an agency and of a law, it is also the biography of a public policy. In this case, from the perspective of policy history, it goes from initial ideas to final burial (at Interior) and rebirth (in Commerce). It is a full circle of the public policy process, or at least a complete chapter of this particular policy history. In the longer perspective that history offers observers of public policy, USTB's eventual extinction is also the story of a policy failure. Zelizer has argued that the historical examination of failures can be as interesting and instructive as successes. He noted that the contemporary revival of political history needed to include forgotten events and specifically failures (2012). Recounting these stories can be as valuable and helpful as the remembered narrative of major events and policies that were enacted and remained in force.

For historians, the saga of the USTB fits into several disciplinary subcategories, including the presidency and Congress, FDR and the New Deal, federal executive-branch expansion, and twentieth-century American history. Smaller parts of the story entail the records of Presidents Truman, Johnson, and Nixon (chap. 9).

For the study of public relations and mass communication, the record of USTB explores a relatively novel manifestation of the practice of public relations in the American public sector. There is a growing literature on the governmental promotion of tourism by other countries, including Canada (Dawson 2004; Dubinsky 1999), France (Furlough 1998), Great Britain (Buzard 2001), and Nazi Germany (Baranowski 2004; Semmens 2005). Generally, the historical literature of American public relations focuses on the for-profit business sector and, sometimes, on the nonprofit sector (aka NGOs). Knowledge of the history of the practice of PR in public administration has been relatively limited (Lee, 2014; 2017a; Lee, Likely, and Valin 2017). There is also an emerging literature on place branding by governmental entities as a form of marketing, tourism destination promotion, and local identity (Zavattaro 2015; 2018).

This is the story of a government agency whose sole assignment was to engage in external communication. Generally, government PR was a controversial subject, especially with the congressional conservative coalition and particularly when FDR was president. Yet here was a rare example of an executive-branch agency assigned by Congress to engage in public relations—and just public relations—for a goal that was declared through the law-making process as in the national interest. Using PR to increase domestic tourism was, unusually, suddenly declared a good thing for government to do. This then justified the expenditure of tax dollars and the employment of civil servants. That FDR was president (and, in fact, in the midst of running for an unprecedented third term) did not seem to bother these conservative lawmakers. Once up and running, USTB become something of the government PR equivalent of a retail big-box category killer or of a regional shopping mall. It engaged in just about every mass-communication technique then available to reach the public at large. It sought to persuade citizens of the value of traveling to see other parts of the country. USTB was government PR to the max.

In the field of African American studies, USTB was something of a quiet standout in FDR's administration regarding his civil rights record. It made a distinct and positive contribution by viewing African American citizens as equally important potential tourists as white Americans. USTB disregarded possible backlash from attacks by racist politicians in Congress. It hired three African Americans to work in its New York office, produced marketing materials to encourage African Americans to travel, printed guides with locations of hotels and guest houses that welcomed African Americans, and included a regular column addressing the interests of African American travelers in the newsletter that USTB mailed to travel professionals throughout the eastern United States. It also openly cooperated with the privately published *Green Book* guides for African American tourists.

A relatively new and growing academic discipline in American institutions of higher education encompasses tourism, travel, leisure, parks, and recreation. This book is about the earliest history of tourism and park promotion in the federal government, including the first federal agency formally tasked with encouraging domestic tourism. Its story also reflects the emergence of the travel industry as a major economic sector and the politics of tourism promotion. Given that USTB presents a template (or at least a chapter) regarding federal involvement in travel promotion, it may well suggest outlines of future similar struggles over this federal policy role, including political debates about the sources and levels of funding for such programming.

Finally, the history of USTB reflects the early stirrings of environmental protection as a field of study. In its initial years, as a unit within the National Park Service, USTB encouraged Americans to visit parks and thereby gain an appreciation for the vistas and unique outdoor sites that NPS was protecting. In particular, as president, FDR had significantly expanded the amount of federal lands protected from private ownership in order to enhance the conservation role of the national government (Brinkley 2016). USTB's promotion in the late 1930s and early 1940s of domestic tourism was a way to encourage Americans to visit these newly protected areas, to appreciate the importance of conservation, and to generate support for

Roosevelt's accomplishments in this matter. This environmental orientation of travel was further emphasized by USTB's post–WWII incarnations. As the US Travel Division in 1946–49, it explicitly promoted conservation and the prevention of stream pollution as impetuses for increasing domestic tourism (chap. 8). USTB's third existence, as NPS's Travel Office during the final years of the Johnson administration and initial years of the Nixon presidency, coincided with the rise of the modern environmental movement, such as the first Earth Day and the establishment of the Environmental Protection Agency (EPA). During those years, the head of the office linked the rise of interest in the environment with travel so that Americans could see for themselves important examples of NPS's protection of the physical and natural environment (chap. 9).

Research Methodology and Structure

In 2018, the editors of *Public Administration Review* noted that triangulation "seems to have been rediscovered by public administration scholars in recent years" (Battaglio and Hall, 825). As a research design, triangulation is an exceptionally useful approach for reconstructing the historical record of an institution, particularly because it helps fill in lacuna of any individual sources (McNabb 2018, 46, 289, 379, 417–18). Its benefit is "piecing together many pieces of a complex puzzle into a coherent whole" (Jick 1979, 608). Another advantage is the ability to create "a chronological reconstruction" of events and developments (van Thiel 2014, 149). Finally, triangulation can sometimes corroborate or contradict accounts from another source. When coming from a second unrelated and independent source, this adds credibility and veracity to events and developments. It is also very helpful in identifying discrepancies about those events. For this study, the three sources I relied on were archival documents, official federal publications, and contemporaneous journalism (Lee 2019).

An obstacle to reconstructing USTB's history was the relatively limited amount of archival material on the agency at the National Archives II site in College Park (MD). Other scattered archival sources (acknowledged in the preface and listed individually in the bibliography) helped fill in many gaps. However, one notable exception of missing documentation was the Interior Department's Secretary's Orders. These legal documents are rarely cited in historical and public administration literature, even though they are the cabinet-level counterparts to presidential executive orders. These secretarial orders were authoritative decisions, promulgated as formal legal documents, and were usually numbered. Several important administrative developments regarding USTB occurred through secretary's orders. However, they were not in "Numbered Secretarial Orders, 1925–1967" of the Records of the Secretary of Interior (RG 48) at National Archives II. Similarly, they were not in the reference collection of the Department of Interior Library,[4] the Departmental Orders folder in Box 156 of the Ickes Papers at the Library of Congress Manuscript Division, nor were they published in the *Federal Register* (as some later ones were). I also consulted with previous NPS historians[5] and the government information librarian at the Wisconsin Historical Society.[6] The only relevant secretarial order I found was Secretary Udall's 1968 decision to revive the bureau, located in the papers of NPS director Hartzog.

Other helpful sources for reconstructing USTB's story were official federal publications. These included congressional hearings, committee reports, presidential messages, and committee documents. In one case, an unpublished congressional hearing was in the ProQuest congressional database. The Travel Bureau's publications were similarly helpful. Some were serials, such as its *Official Bulletin* and the newsletters of its New York and San Francisco field offices. The bureau also occasionally issued other publications, such as calendars of events and research reports.

Even though journalism is sometimes discounted by academic historians as a nonoriginal and nonprimary source, it can be very valuable in the triangulation approach as an independent source that might confirm or supplement other sources. Such contemporaneous material has another particular benefit of presenting a "you are there" perspective. Reporters and other writers were conveying how things looked at that point, with no idea how the story would turn out in the long run. This provides a good counterbalance to the inevitable tug of hindsight that often invisibly plagues historians. There is an unintended tendency to disregard or downgrade that which seems ultimately unimportant based on later developments. The arc of historical storytelling of governmental history should be about letting the story unfold as it did, particularly when the outcome of events was uncertain or unpredictable. From that perspective, historians need to recognize that journalism is indeed "the first rough draft of history."[7]

When it was necessary to interpret and analyze the historical record or connect the dots due to gaps in source material, I applied the historical evaluation standard Clark used for identifying the origins of World War I. He called it the logic of "maximum plausibility" for explicating events and motivations (2013, 48). This approach can somewhat strain the justifiable academic insistence on careful judicious documentation. However, in a few cases, it seemed necessary to try to pull back the curtain and examine possible motivations and political considerations. I have been as restrained as possible about using Clark's historiographic technique.

A note on nomenclature. The federal agency promoting domestic tourism went through multiple name changes. It began as the US Tourist Bureau; then it became the US Travel Bureau. After WWII, it was US Travel Division and in the late 1960s and early 1970s, its name was the Office of Travel and Information Services. During its existence, the agency's longest lasting title was US Travel Bureau. Therefore, for simplicity sake, when discussing the overall record of the domestic tourism promotion effort at the NPS and the Interior Department, I have used the name US Travel Bureau and its acronym (USTB) as a generic term. This covered not only when that was its formal and legal title, but also in other post-WWII contexts as an overall reference to its activities throughout its existence.

The structure to follow is chronological, beginning with the pre-FDR congressional efforts to involve the federal government in tourist promotion (chap. 1), the administrative creation and operations of the US Tourist (later Travel) Bureau (chaps. 2–4), the congressional act creating USTB as a statutory agency in 1940 (chap. 5), and the programs of USTB until Pearl Harbor in December 1941 (chap. 6). USTB gradually went out of existence after the US declared war, lasting in reduced form until early 1943 (chap. 7). With the end of the war, the Truman administration tried to revive it, but it shut down again in 1949 due to Congress refusing to fund it further (chap. 8). Then President Johnson's administration revived it in 1968–69 and Congress funded it through 1971. The Nixon administration was not as interested and declined to propose funding USTB for FY 1972. Then, in late 1973, President Nixon terminated USTB's legal status by signing legislation transferring the statutory duty for promoting domestic travel from the Interior Department to the Commerce Department (chap. 9).

1

The Tourism Industry and Big Government
Origins of the US Travel Bureau, 1930–1936

Republican Origins: Seventy-First and Seventy-Second Congresses, 1930–1933

The idea that the federal government should play an active role in promoting tourism predated the New Deal. In 1930, during the Republican-majority Seventy-First Congress, Congressman Leonidas Dyer (R-MO) introduced a bill to create a Division of Travel within the Commerce Department's Bureau of Foreign and Domestic Commerce.[1] He emphasized the value of promoting visits to the sites of the National Park Service (NPS), whether to see the natural landscapes or to learn more about the nation's history. He summarized some of the major attractions of the parks and sites managed by the NPS. Increasing visits to parks would not only be patriotic, it would also be good for the many businesses relying on tourism and travel.[2] The next year, he pushed for enactment of a second version of the bill, this one formally endorsed by the newly organized International Travel Federation, which was a unit of the US Chamber of Commerce.[3] Dyer emphasized that the US economy would benefit from increased foreign tourism and that promotion abroad would assist in doing so, similar to what other countries were doing.[4]

The House Interstate and Foreign Commerce Committee held a public hearing on the bill in January 1931 (US House 1931). By now, the economic interests were relatively well organized and spoke about the benefit of the bill to various segments of the business sector that would benefit directly from increased tourism—whether from abroad or domestically. Testimony came from two representatives of the International Trade Federation and a spokesman for the American Travel Development Association. They discussed the size of the travel sector of the economy and the potential economic benefits to these businesses if the federal government were to engage in tourism promotion. Most of the hearing was dominated by three civil servants from the Department of Commerce's Bureau of Foreign and Domestic Commerce. They spoke generally of the economic benefits that increased travel could have and released results of an in-house research report on what other national governments did to stimulate tourism to their countries (Bratter 1931). The idea seemed to be that, if only to keep up with the Jones's, this bill was a logical idea.

They treaded a careful policy, ideological, and political line. The federal government should not do what private businesses were doing. Rather, "only do those things that individuals and organizations for touring purposes can not do themselves because of their lack of facilities" (US House 1931, 2). It would serve as an information clearinghouse, a coordinator and liaison with American businesses, and could add tourism promotion to the responsibilities for the Department's extant overseas commercial attachés, but not to the staffing in any of the department's domestic regional offices.

However, the department (and, by implication, President Hoover's administration) did not endorse the bill; it was noncommittal on it. The general statutory mission of the department could be interpreted as already permitting these new activities. While the legislation might not be necessary in terms of authorizing these particular activities, some additional appropriated funds and staffing could be helpful, but the department was not *asking* for it. This position on the bill permitted the administration to avoid attacks that it was requesting an increase in federal spending (and powers) in the midst of the Great Depression. Given that President Hoover consistently opposed having the federal government get involved in direct relief payments to the unemployed, Congressman Sam Rayburn (D-TX) couldn't resist the political observation that the bill would constitute "administrative relief," and therefore—if endorsed by the administration—would be a political double standard (p. 4).

Nothing happened after the hearing. Dyer kept pushing, introducing a similar bill in the next Congress (Seventy-Second) in December 1931.[5] For that Congress, which had been elected in November 1930, the majorityship shifted to the Democrats. A bill sought by a minority party member and endorsed by a unit of the Chamber of Commerce did not have any political attractiveness to the Democrats. In January 1933, just before the final adjournment of the Seventy-Second Congress, Dyer introduced yet another version of the bill.[6] By now, the 1932 presidential election had already occurred, and the Democratic Congress did not want to do anything that would tie the hands of President-elect Franklin Roosevelt or deny him the political credit for proposing any new programs.

Democratic Initiatives: Seventy-Fourth Congress, 1935–36

The momentous Hundred Days at the beginning of FDR's presidency and the subsequent flurry of legislative enactments occurred during the Seventy-Third Congress. The central preoccupation was on dealing with the Great Depression, especially key problems of unemployment, relief, farm prices, property foreclosures, and the financial system. Secondary issues would have to wait. Also, notwithstanding emergency spending bills, the administration presented itself as fiscally conservative, seeking to accomplish a balanced budget (Zelizer 2012, chap. 7).

The issue came back two years later during the Seventy-Fourth Congress. In 1935, similar bills were introduced by senior members of the majority party from states heavily benefiting from tourism. Senator Royal Copeland (D-NY) chaired of the Commerce Committee and Representative Clarence Lea (D-CA) was third ranking Democrat on the Interstate and Foreign Commerce Committee.[7]

The Senate bill proposed creating an independent commission in the executive branch called the US Travel Commission. The Commission would have two members, the Secretaries of State and Commerce, and could employ its own staff. At a Senate hearing in May 1935, public testimony was in support of the bill, including the American Hotel Association (US Senate 1935a). However, bureaucratic politics and the Administration's spending priorities quickly came to the fore. Secretary of Interior Harold Ickes wrote that he generally supported the idea (even though his department would have no representation on the Commission). Parallel to Copeland's bill, Ickes had prepared a bill draft placing a new Travel Division in Interior's National Park Service. However, the Bureau of the Budget—using its legislative clearance role—informed him that the idea "would not be in accordance with the financial program of the President" (p. 3). Hence, he could not endorse Copeland's bill. Copeland feared that this "adverse report" from BOB would fatally harm the chances of passing any travel legislation during the 74th Congress.[8]

Secretary of State Cordell Hull was sympathetic to the general policy goal, but was compelled (by BOB's position) to conform to the administration's austerity effort "to reduce Government expenditures within the limits of activities which are vitally necessary to the welfare of our country," i.e. emergency spending programs to deal with domestic unemployment. Therefore, he could not support creating a new government agency (pp. 3–4).[9] A representative of the Commerce Department also stated the Department's opposition to the concept of an interdepartmental commission and expressed a preference that if any bill were to pass, it should assign the tourism duty to its Bureau of Foreign and Domestic Commerce.

A month later, a subcommittee of the House Committee on Interstate and Foreign Commerce that was chaired by Lea held a hearing on his bill (US House 1935a). It proposed creating a Tourist Travel Division in the Commerce Department's Bureau of Foreign and Domestic Commerce. Testimony in support came from the American Steamship Owners' Association (pp. 20–21), as well as a repeat of the hotel association's Senate testimony (pp. 9–20). At the end of the hearing, two letters from the Department of Commerce were inserted into the record. The director of the Bureau of Foreign and Domestic Commerce stated in a memo to the secretary that he had "no objection to the Department approving it in principle." However, the cover letter from Commerce Secretary Daniel C. Roper to Committee Chair Sam Rayburn said that Department opposed the bill because it "would not be in accord with the financial program of the President" (p. 22).

Hinting at the changing political landscape, over the summer both committees recommended adoption of their respective bills, notwithstanding the opposition of the administration of their party. The committees recommended the bills without amendments and, apparently, without any opposition (as there were no minority reports). This indicated changes since the previous consideration of the idea in the early 1930s. There was the potential of a bipartisan consensus of the benefits of promoting tourism as a form of economic stimulus, as well as of the travel business becoming increasing well-organized as a special interest group in the capital. Using taxpayers' money to benefit their industry sounded like a great idea. This was further indicated by supportive coverage in news outlets oriented to business. *Business Week* reported favorably on the Senate bill, noting "Washington's late recognition that the United States is losing a big opportunity to develop a huge and profitable tourist business."¹⁰ Similarly, *Traffic World* told its transportation industry readers of the "favorable" Senate Committee report on the bill and reprinted the report verbatim.¹¹

For the most part, the Senate Commerce Committee's report stuck to the general arguments for encouraging foreign and domestic tourism and avoided addressing bureaucratic turf battles and the administration's opposition on spending grounds (US Senate 1935b). A few weeks later, the House Committee on Interstate and Foreign Commerce also recommended adoption of the House version of the bill. After summarizing the case for the bill, the report reprinted without comment Roper's cover letter opposing it on general financial grounds along with the memo supporting it in principle from the head of Bureau of Foreign and Domestic Commerce (US House 1935b).

The bill first came up for consideration on the Senate floor over the summer. It began inauspiciously. A Republican asked that debate be postponed because a fellow Republican was opposed to the bill but was absent that day. Copeland claimed this was a misunderstanding, that the missing member was in favor. Then, Majority Leader Joseph Robinson (D-AR) wondered out loud if the federal government should get into the advertising business. That was enough to postpone consideration.¹² But, a few weeks later, it came up again and passed with almost no debate and no roll call vote.¹³

In early 1936, the *New York Times* reported that Congress was getting more interested in "a better balance in the tourist trade."¹⁴ With the Senate bill now in the House, Congressman Lea tried to short-circuit the legislative process by calling up the Senate bill to the floor for passage without routine consideration by the appropriate House committee. Revealing how the politics of the bill had been turned upside down compared to the pre-FDR presidency, Republicans now objected to bringing it up notwithstanding business support for the bill. It was enough for them that majority Democrats wanted to pass it and that FDR would get yet another expansion of federal activities. Congressman James Wadsworth Jr. (R-NY) objected because the bill would be "the beginning of a new bureau in this Government."¹⁵ A few months later, Lea tried a different tactic, bringing his House bill to the floor given that the committee had recommended it favorably. First, a Democratic member, Thomas O'Malley (WI), argued against the bill, saying that it would open the gates to Europeans who intended to stay and would take jobs away from American citizens. He felt that, given the state of the economy, even with a tourist bureau, few Europeans would have enough money to travel to the United States. He also believed that Europe should pay its war debts to the United States first. James Taylor (R-TN) argued against it because it was a subsidy, something he opposed, even if business was for it. A different Republican, Leroy Marshall (R-OH) objected to calling up the bill, ending consideration.[16]

These legislative shortcuts all blocked, Lea tried the more traditional route of a committee report on the Senate bill. In March 1936, the House Interstate and Foreign Commerce Committee recommended passage of the Senate bill but with a substitute amendment replacing the language of the Senate bill with Lea's. The report noted the federal government's "startling lack of cooperative effort" in promoting domestic and foreign tourism and emphasized how the benefits of such advocacy would go to a wide range of businesses, including "transportation agencies, hotels, restaurants, [and] retail establishments" (US House 1936, 3–4). That strategy did not change the new upside-down partisan politics of the bill. Republicans again objected to two efforts to bring the bill to the floor during the spring.[17] Given that it was an election year (and with Congress adjourning sine die on June 20, 1936), their objections essentially killed the idea for the rest of the Seventy-Fourth Congress.

By now, the administration was in favor of promoting tourism in principle, but hoping to avoid these political fights on Capitol Hill, with their resultant accusations of more bureaucracy, big government, and increased deficit spending. In 1935, Secretary of Commerce Roper convened a conference in Washington of travel-related businesses and called for "a systematic program" to promote tourism.[18] This led to the creation of the National Resorts and Parks Association, with former US Ambassador to Germany James Gerard as honorary president. President Roosevelt sent him a congratulatory letter and noted the benefits of increased domestic and foreign tourism.[19] The association promptly opened an office in Rockefeller Center's new International Building (McGarry 1936).

2

The US Tourist Bureau

Birth by Administrative Action, 1937

Who Needs Congress?

Lurking behind the scenes during the 1936 maneuverings on Capitol Hill was the bureaucratic turf battle over where any future federal travel bureau should be placed. The Commerce Department insisted this was within its mission, and the various legislative proposals in 1936 confirmed that. After all, the travel business was a component of interstate and foreign commerce, a federal role specified in the US Constitution. Interior Secretary Ickes thought otherwise, especially if the goal was to increase visits to national parks—a theme near and dear to the president's heart (Brinkley 2016). A later departmental publication asserted jurisdiction based on the rationale that "touring is fundamentally recreational. Recreation is an important factor in the conservation of the Nation's human resources, and the Department of the Interior is the conservation department of the Government. Furthermore, the National Park Service, in that Department, is the recreation agency of the Government" (US NPS 1938, 33).

The first months of 1937 were a time for new beginnings. Roosevelt had been re-elected in November 1936 by a surprisingly strong margin. Therefore, the start of his second term in January was a time for new initiatives by the administration. Ickes was eager to do his part. Before the new Seventy-Fifth Congress would convene on January 5, 1937, it was open to anyone to make the first move. Ickes at Interior did. The day before the new Congress was sworn in, Arthur E. Demaray, NPS's acting director, submitted to Ickes a formal request to create a "United States Tourist Information Bureau" within the agency.[1] In mid-1936, Congress had passed a law directing NPS to conduct a comprehensive survey of recreational facilities provided to the public by all levels of government.[2] Based on the data collection and interactions NPS subsequently had with state parks departments, it now possessed a detailed catalog of the travel promotional efforts of these state agencies. As a result, NPS and Interior now had a leg up on the race for a federal travel bureau. Interior had a *list*, and it was *park* oriented. This strengthened a framing of the argument that travel promotion should be based in NPS, even if it would be oriented to encouraging domestic tourism in general (and international tourism, too). In his proposal to Ickes, one of Demaray's explicit rationales was that creating a *fait accompli* might influence future congressional proposals toward specifying that a statutory tourist bureau be in Interior instead of, as in the 1935–36 versions, Commerce. Another rationale he made for administrative action was that emergency relief programs could be used to jumpstart the new bureau. In particular, WPA writers and artists could generate material for the new bureau as well staffers funded by the Civilian Conservation Corps (CCC) from the Emergency Conservation Work (ECW) Act of 1933. Demaray proposed that Nelson W. Loomis, at the time an associate recreational planner who had worked on the national study (which had been funded by ECW), be appointed to head the new organization and that it begin with a single public office in New York City.[3]

On February 4, 1937 (before the new Congress had time to act on any tourism legislation), Ickes approved establishing the Tourist Information Bureau within NPS. The department quickly issued a press release announcing it a week later.[4] The news was trumpeted in the press, especially the Sunday travel sections and sports pages (the latter regarding outdoor recreation). The outdoor column of the (conservative) *Los Angeles Times* lauded the idea and noted that the WPA Writers Project was preparing some travel guides for free distribution.[5] The *Charleston [WV] Gazette*'s outdoor column mentioned that the state's Conservation Commission quickly notified travel-related groups around the state to send copies of their pamphlets and brochures to the Bureau's New York office for use when responding to citizen requests for information.[6] By late March, the *New York Times* reported that the five employees of the Bureau had already "received hundreds of inquiries" and that Ickes "hopes to develop the tourist services along the lines similar to those maintained by foreign governments."[7] More coverage occurred over the summer, with two national press services distributing articles about it and another short piece in the Sunday travel section of the *New York Times*.⁸

Trying to embed the office more deeply within Interior's portfolio, a departmental press release in April noted that the NPS's "Tourist Bureau" (now capitalized and with a shorter title) would also disseminate information about travel destinations outside the forty-eight states, specifically the territories administered by the department, such as Alaska, Hawai'i, Puerto Rico, and the Virgin Islands.⁹

Operations and Public Relations, 1937

Interior and NPS moved very quickly to make the travel bureau a reality and ongoing operation. As planned, funding came from allocations of emergency appropriations to CCC and WPA (US House 1938, 538–39). The bureau's five staffers had already been involved in travel and recreation work, such as for NPS's national park and recreation survey. The New York office was initially located on the fifteenth floor of the federal courthouse in Manhattan. However, it quickly became clear that this venue was not amenable to drop-by visits from citizens seeking travel information. Loomis promptly moved the bureau to the ground floor of a federal building at 45 Broadway.¹⁰ The area was informally known as Steamship Row for all the ticketing offices of passenger shipping lines there.

Loomis felt that the new agency needed increased visibility and gravitas. This was confirmed after he gave a talk at a travel industry conference. A newspaper in Tennessee welcomed the creation of the new agency but confessed that its establishment "had escaped public attention until a few days ago" when Loomis spoke to the conference.¹¹ Therefore, Loomis proposed that Secretary Ickes appoint Gerard to an uncompensated position in the USTB, with the public administration title of collaborator.¹² Gerard's role would be as a kind of honorary director of the bureau and particularly could use his high public profile to promote travel to the United States from Europe and Latin America.¹³ Ickes approved it, and the appointment was announced in September 1937.

The news seemed to make the bureau instantly important. Gerard's appointment garnered extensive publicity, an indication of his standing as a noncontroversial and above-politics elder statesman and public figure.¹⁴ Gerard promptly dove in, giving radio addresses to promote the bureau. On

October 5, he spoke on WOR on *See America First*. A week later, he appeared on the Columbia national network. In that broadcast, he praised the new federal travel initiative, focusing on the economic benefits that would accrue to Americans throughout the country, including the patriotic role of visiting national parks, the potential for increasing foreign tourism, and how tourists from Latin America could strengthen FDR's Good Neighbor policy. He also assured listeners that the Tourist Bureau would not be competing with the private sector's promotion of travel or for-profit travel agents. Rather, this new government agency would have a coordinating and cooperating role with business.[15] However, Gerard's role as USTB's collaborator was more than symbolic. For example, in various formal Bureau publications, following the standard hierarchal listing of the name of the departmental secretary and the head of NPS, Gerard was listed, not Loomis (USTB 1937a, 1937b). The travel editor for the *New York Herald Tribune* was fooled, referring to Gerard as "head" of the bureau and not mentioning Loomis at all.[16]

The benefits of all this publicity then began showing up in editorials praising Gerard's efforts to promote domestic and foreign tourism. This hinted at the unusual political opening that USTB and Gerard had created. Positive editorials about a New Deal program were few and far between. A Massachusetts newspaper doubted that, due to European tensions, there was much potential for foreigners to come to the United States, but praised the idea of Americans getting better acquainted with their country. This was "worthy material" for Gerard to promote.[17] A newspaper in Texas praised Gerard and the new bureau's efforts because of their "economic desirability . . . and the dollars involved."[18] An editorial in an Illinois newspaper endorsed Gerard's efforts as "a felicitous gain of good will," whether by foreigners to the United States or citizens for their country.[19] Reverting more closely to type, unable to see any good in any New Deal program, a different Massachusetts paper sneered that this was yet another new federal agency, embodied an ideology believing government could do something better than business could, and that tourism did not need an "added stimulus" from Washington.[20]

During the second half of 1937, USTB gradually institutionalized itself into an organized and full-service government information bureau. The office on Broadway included street-level window displays and other exhibit cases from NPS's Museum Division. Inside, it had a long service counter and display racks with brochures printed by governmental and for-profit entities promoting travel. Similarly, the mail brought hundreds of requests for information. These queries were answered by mailing back the appropriate printed material in the bureau's inventory. Replying to these individual requests covered about half the volume of work. The other half involved interacting with travel-related businesses, attending travel industry conferences, proactive information dissemination, and ongoing contacts with USTB's partners.[21] For August, September, and October, the Bureau reported receiving about 6,000 requests for information, including 1,800 in-person visitors, 350 telephone inquiries, 600 requests for information by domestic mail, plus 100 from foreign citizens. At an auto show in Manhattan, 3,300 people stopped at the bureau's booth.[22]

The bureau's organizational structure consisted of two line units, the Information Section and the Publicity Section. The purpose of the information unit was to disseminate information to the public, presumably on a retail one-to-one basis. However, that the other of USTB's two line units was for *publicity* demonstrated how central public relations was to the work of the Bureau. This did not seem to be a problem with legislators who otherwise were generally quick to condemn New Deal agency propaganda. USTB also had a third section, Operations, which handled administrative tasks involved in managing any government entity, such as the generic functions of budgeting, HR, and supplies. An assistant supervisor headed each of the three sections.[23]

The new bureau had its inevitable glitches and growing pains. At one point, Loomis had to apologize to a higher-up at NPS, confessing that he was "somewhat new to Government routine and do appreciate the valuable pointers you have given me."[24] The lack of direct congressional appropriation also hampered the bureau's growth, having to scrounge and scrimp to cover its budget. In particular, the ability to print its own materials (as opposed to distributing publications by state travel and park offices) was under severe congressional restrictions. In its annual report for the fiscal year ending in mid-1937, NPS lamented the tight congressional controls on printing budgets that limited the ability of its new travel bureau to "spread the gospel of national parks abroad and at home" (Ickes 1937, 70).[25] Slightly offsetting that, American Express donated $3,500 to extend the operations of the bureau and committed to cooperating with it.[26] There was also some confusion in the press about the bureau, with a reporter stating that it had a second office providing service to the public in Washington.[27] That was inaccurate. The bureau's only public contact location at that time was the field office in New York.

President Roosevelt gave an indirect boost to the mission of the bureau and the administration's encouragement of domestic travel. In an October national radio address, ostensibly as greetings to the annual civic forum of the *New York Herald Tribune* (a Republican newspaper and FDR critic), he called for "an intensive drive to get people to know their own country better." Lightheartedly, he amended Horace Greeley's famous advice of going west to include "all the way to the [Pacific] Coast—[and] should go South and North and East," as well (Roosevelt 1969, 1937: 413).[28]

∽

Ickes had reason to be happy with what he had accomplished in 1937. But he was *not*. For that, he would need Congress to pass a permanent law creating USTB and placing it in Interior. He would also need the president's approval. It would be a long battle to get there. In the meantime, his goal was to have the new agency up and running. New political coalitions could create facts on the ground that could not then be undone, or at least not easily.

3

The US Travel Bureau

Renamed and Expanded, 1938

Organization and Leadership

Although still limited to funding from emergency appropriations such as WPA, the bureau continued institutionalizing itself in 1938. In March, NPS's Arno Cammerer approved a reorganization giving assistant NPS director Conrad Wirth direct line supervision over the bureau. Somewhat oddly, the New York field office would "report directly" to Wirth instead of through Loomis.[1] The New York office then began submitting weekly reports to him on its activities.[2]

A few months later, Loomis was given the title of bureau chief and was in charge of the bureau's Washington headquarters office.[3] It contained a new reference library of travel-related materials; an editorial section to write material for publications, press releases, and radio scripts; a research section; and an arts section to create exhibits and illustrated material.[4] The bureau's Washington office also began welcoming walk-in visitors and tourists seeking information.[5]

In June, the bureau somewhat expanded its scope by changing its name from US *Tourist* Bureau to US *Travel* Bureau. The rationale was that "the term 'tourist,' as popularly used, does not include all travelers who would benefit from the bureau's services."[6] Agency staff were a bit miffed when a newspaper article on the bureau a few weeks later referred to it by its old name.[7] (Inevitably, there's always someone who doesn't get the word. Two years later, an editorial in the *Boston Globe* still referred to it as the US

Tourist Bureau.[8]) Also that month, Gerard resigned as USTB collaborator. In a letter to FDR, he said that this official status (even though unpaid) "cramps my status in raising funds for some of our friends,—a task that as usual I have been asked to take up."[9]

The New York office reported it was doing "a land-office business" and had added fifteen additional staffers in the second half of the year to handle the volume of requests and visitors.[10] USTB then opened its next field office in San Francisco in September, partly in anticipation of greater travel due to the upcoming world's fair in San Francisco in 1939–40. It, too, was "equally busy" as the New York office.[11] J. L. Bossemeyer,[12] head of the San Francisco office, began giving talks to business and civic groups about USTB's efforts to promote the tourism business on the West Coast.[13] Along with the overall bureau mission to promote travel generally, the new office would particularly focus on travel pertaining to western destinations. Loomis also spoke at a western conference to report on the bureau's efforts to promote visits to the fair.[14] It adopted the slogan of "See All the West in '39"[15] and began distributing information about skiing locations in the West.[16]

While western states were reflexively conservative and hostile to the feds (in part because it controlled enormous swaths of land), they were also early in recognizing the economic benefits of tourism. The age of automobiles had facilitated a new form of travel and spending, evocatively called "windshield wilderness" (Louter 2001). Indicating the political opening that USTB gave the administration with business in the American West, Loomis was invited to address a conference of eight state chambers of commerce in Salt Lake City. The purpose of the conference was "coordination of the region's bid for tourist travel" along with the West's traditional probusiness and anti-Washington goals of "protecting mining interests, and for promoting the welfare of grazing, irrigation and other industries."[17] Montana's travel office also posted a staffer at the bureau's New York office to promote its tourist spots to easterners, including dude ranches.[18]

Careful not to offend or downplay other regions of the country, the bureau also had tentative plans to open additional field offices in Chicago and New Orleans.[19] (It never happened.) Demonstrating support for travel to all regions and localities, USTB published a brochure with travel routes in the South, and Loomis spoke at a recreation and park conference in Atlanta.[20] An editorial in a Boston newspaper lauded the bureau, particularly because increasing tourism would "be of great value to New England."[21] In another geographical effort, in cooperation with the government of Puerto

Rico (then under the jurisdiction of the Interior Department), it organized a cruise to the island. As hoped, the cruise prompted a press coverage that lauded the attractiveness of Puerto Rico as a tourism destination.[22] Further promoting US territories in the Caribbean, the bureau also encouraged visits to the US Virgin Islands (also in Interior's portfolio).[23]

USTB's new research arm released its first report in 1938, calculating the benefits that would accrue to the auto industry from increased travel (Dorsett 1938). The bureau also prepared a study of the impact of the travel industry on tax revenues,[24] released several economic updates with the most recent travel data,[25] compiled a directory of state travel agencies (USTB 1938a),[26] and produced a listing of businesses and organizations involved in camping (USTB 1938b). Other efforts to embed the bureau into the travel business included a trip by Loomis to address the annual convention of the American Hotel Association in Galveston (TX) in September[27] and a conference with leaders of the tourism industry convened by Interior Assistant Secretary Oscar Chapman in December at Interior's headquarters in Washington.[28]

Public Relations

In particular, USTB's publicity and PR activities gained significant momentum in 1938. It put major emphasis on utilizing free radio airtime to promote travel. The year kicked off with a nationwide broadcast by Interior Secretary Ickes on January 17. He described the services of the bureau and explained the rationale for its existence, including patriotism of getting to know one's country better, increasing foreign tourism, visiting national parks, and implementing President Roosevelt's Good Neighbor policy.[29] Ickes read on the air a letter from FDR praising USTB as "another valuable service" to the citizenry and to promote visits to parks. Softly, but explicitly, the president took advantage of the opportunity to praise not just USTB, but more generally the expansion of the federal government under the New Deal. He said, "One of the most significant trends of our time is the growing concept of government as a social institution kept constantly geared to serve the needs of the people, rather than a static instrumentality for the mere preservation of law and order."[30] These tangible benefits to individual citizens provided by the bureau were but one example showing the advantages of his approach to government and what would be missing if the opponents of the New Deal had their way. Ickes's radio address was also covered as a news event.[31] As it turned out, privately, Ickes was miffed that the president declined to give the speech himself.[32] The probable explanation is that FDR had an intuitive sense of the concept of (what years later was called) overexposure and sought to limit his radio appearances so that they were infrequent enough to seem special and worth tuning in to.

Gerard was on the radio several times promoting USTB, including as a guest on the popular radio program hosted by Lowell Thomas.[33] The bureau sponsored a weekly radio series with travelogues about individual states, with fifteen programs broadcast between January and May, 1938.[34] Loomis had a regular fifteen-minute spot every Friday on New York City's municipal radio station.[35] The head of the New York field office, J. R. Anderson, was the featured guest on a national network series *Your Government at Your Service* in August, and J. L. Bossmeyer, head of the San Francisco office, in November, described the work of USTB on the series *Question Box Program* regularly aired on a local radio station.[36]

Newspapers also continued their generally positive coverage of this New Deal program. Perhaps somewhat reluctantly, even some editorial boards commented favorably about USTB. In some cases, this was motivated by parochialism. A New Orleans paper liked the fact that an increase in tourism from Latin America might mean that the city "is in line for a good share of the business."[37] A Miami newspaper noted that Europeans seemed to have discovered and liked vacationing in sunny and safe Florida in contrast to the upheavals going on in their home countries. "The department of the interior is aiming high . . . but if it succeeds in bringing any considerable amount of foreign travel to this country, its efforts will be justified."[38] Other papers thought USTB was a good idea for the country as a whole, whether to promote patriotism or the economy.[39]

Near the end of the year, USTB got another major radio boost. A radio studio built on the top floor of the Interior Department's new headquarters building in Washington was financed with funds from the Public Works Administration (PWA) and ready for use in late 1938 (Lee 2011, 154–55).[40] Given the routine criticism of agency PR programs (let alone having a federal radio station), Ickes sought to have the inaugural broadcast from the new studio be as noncontroversial as possible and to try to build in some favorable press coverage. The *Washington Star* had been sponsoring a nationally broadcast series called the National Radio Forum and welcomed the privilege of being involved in the dedication of the new facility. This gave the event some political cover because even conservative politicians would be reluctant to criticize the routinely conservative paper. Also, Ickes could count on the paper to give the event maximum coverage. The forum was a panel of government officials from several executive-branch agencies discussing the roles of their agencies in promoting hemispheric travel and tourism. At the end of the broadcast, Ickes put in a plug for congressional approval of a bill to authorize and fund the USTB.[41] After it was over, Ickes privately thought the program was dull and uninteresting,[42] but it of course got good coverage in the *Star*.[43] A full-page article in a travel industry magazine praised the broadcast for promoting travel in general and hemispheric travel in particular.[44] To promote support for USTB, Loomis sent copies of the fourteen-page transcript to leaders in the travel industry. In his cover letter, he said that "Secretary Ickes would welcome your expression of opinion" about the program and that their feedback would help "the Department's plans for the further expansion and development" of USTB.[45]

The bureau continued to be the darling of newspapers, especially by providing editorial material for the Sunday travel and auto sections. USTB's releases helped fill these recurring news holes with copy that was noncontroversial and helpful to readers. Not only was this welcomed by journalists and editors, but it helped draw readers to the (even more important) paid advertising from travel businesses. Given its bureaucratic home, some of its releases put special emphasis on visiting national parks.[46] One of the bureau's PR products was an occasional feature column by Loomis, which newspapers could also use to fill their travel pages.[47] Even conservative Republican papers such as the *New York Herald Tribune* and the *Washington Star* were glad to write frequently about USTB in their Sunday editions.[48] This was demonstration of another political benefit to the administration of creating the travel bureau: it helped FDR break out of the ghetto of negative political coverage to reach readers in other newspaper sections. Beside the travel and auto pages, coverage in unlikely places included the amateur photography column of the *Post*,[49] a laudatory article on USTB's efforts to encourage European tourists to visit in the monthly magazine *Commentator* (Williams 1938),[50] and a mention in the specialized *Journal of Adult Education*.[51]

During 1938, the bureau also institutionalized outreach to its multiple constituencies and stakeholders by launching a monthly periodical with the ungainly title of *Official Bulletin of the United States Travel Bureau, National Park Service*. The inaugural four-page issue came out in October. Articles included a welcoming column on the front page from Associate Director Demaray on the rationale for a federal travel entity, summaries of USTB's activities to promote domestic travel and tourism from abroad, short pieces on efforts by individual states to promote visits, and a calendar of upcoming major events. The first issue included an invitation for recipients to submit news of their activities for inclusion in future issues.[52] Subsequent issues were generally the same length and had the same format. The second issue had another welcoming statement on the first page from Director Cammerer. It also included an invitation that readers could "reprint any of the text herein contained" in their newsletters and other publications, such as those published by state parks agencies, chambers of commerce, and travel industry associations.[53] This was a low-key but shrewd way to extend awareness of the work of the bureau through a kind of force multiplier, thus reaching farther and deeper into the grassroots and local levels of its business and governmental constituencies.

African Americans

One relatively unusual public outreach effort by USTB was to African Americans. They had long identified with the Republican Party, due to Lincoln, Reconstruction, and the postbellum civil rights amendments to the Constitution. In a major political shift, FDR's New Deal coalition included African American voters in urban areas and (contradictorily) racist Southerners. Yet, once elected, FDR was unwilling to show leadership on civil rights matters if only because the political consequences would be to alienate the Southern mandarins who, due to seniority, controlled both houses of Congress and could thus doom all his legislative proposals. Therefore, civil rights was a political nonstarter in the White House. However, farther from the White House, there was a modicum of attention to racial issues. Some New Deal programs channeled funding specifically to benefit African Americans, and some individuals in the administration, including Interior Secretary Harold Ickes, more openly supported civil rights (Sklaroff 2009, 18–25). In general, though, these lesser efforts and the tone at the top amounted to a "glacial pace of change" on race matters (Lucander 2014, 3).

In that context, it is somewhat surprising that USTB openly and vigorously pushed to increase African American travel within a year of its founding. The New York office hired an African American to run its program, published guides for travelers of color, and issued press releases to minority newspapers. This effort started in 1938 when the bureau hired Charles A. R. McDowell as a staffer with the title of collaborator.[54] That his title was the same as Gerard's, who was unpaid, was perhaps a deliberate way to avoid attention and attacks. The title implied he had a fuzzy personnel status with the bureau, perhaps not even being a formal employee. Despite the misdirection, he was a paid staffer and had "a well-appointed office" of his own at the New York field office, including two secretaries (also African American).[55] About ten years earlier, McDowell had opened a travel agency in Harlem and became the "first person of his race to become a Greyhound Bus company agent."[56] The Urban League had recommended him to USTB. Using language common at the time, one article said that "the appointment of a Race man" would help "to better inform Race travelers."[57] McDowell's goals were to promote the travel business "in all large centers of Race population, thus opening up a hitherto uncultivated field of economic opportunity," encourage more visits to national parks by African Americans, prepare literature "relating to phrases of Race travel," and perhaps produce "an all-Race motion picture travelogue."[58]

That this was a touchy subject was alluded to when one article said McDowell would focus on "problems attendant upon Negro travel throughout the United States."[59] A columnist for an African American weekly stated it more bluntly: "Race prejudice and discrimination take much of the pleasure out of such journeys and being thus conditioned Negroes often fail to take advantage of the facilities offered by the national park service and recreational areas. Unlike the white traveler, the average Negro who ventures forth has many more problems than getting money to go. He wants to know where to stay, what he can do and where he can eat."[60]

Legislative Maneuvering: Seventy-Fifth Congress

The creation of USTB by administrative action in 1937 was part of a larger effort by Ickes and Interior to assert its claim of bureaucratic primacy for travel-related programming. Beside the *fait accompli* and attendant publicity over its initial existence, beginning in late 1937, Gerard and others worked to create grassroots support from the industry for the bureau, implicitly manipulating the travel business to support Interior's claim of priority over Commerce or State. When Gerard and an NPS official addressed a conference of travel-related businesses in September 1937, they called for expressions of support for the legislation they were drafting.[61] Gerard also sent a mass mailing to several thousand opinion leaders asking for their feedback on the proposal to create a federal agency to promote travel. By December 1937, 1,500 had replied, stating their support for such an initiative. Misleadingly,

Ickes issued a press release that this grassroots support from business was specifically in favor of *his* preference to base that effort in Interior's NPS.[62]

Senator Copeland (D-NY), who had sponsored legislation previously, initially sought to finesse taking sides by introducing a bill in November 1937 to create a federal travel commission consisting of three cabinet secretaries: interior, commerce, and state.[63] His bill was promptly endorsed by a newspaper in Massachusetts: "We hope that the Copeland bill will be passed . . . Americans would profit handsomely, and it would be a boon to international goodwill and understanding."[64] But a few months later, he had a change of heart and introduced a new bill giving some jurisdiction to Commerce, but tilted the focus to Interior. The bill directed Ickes to appoint an advisory United States Travel Board. It would have as members not just representatives from Commerce and State, but also as many members from the business sector as Ickes may wish (and they would serve at his pleasure).[65]

As chair of the Senate's Commerce Committee, Copeland skipped holding a public hearing on the bill, feeling it was unnecessary because "this bill has passed the Senate in previous years" (US Senate 1938, 1). In April, the members of the committee voted to recommend that the Senate pass the bill notwithstanding a letter from the secretary of state opposing it because the bill "would not be in accord with the program of the President" (3).[66] Even the Interior Department, so eager for the bill to pass, had to confess in its formal letter to Chairman Copeland on the bill that this bill still had an adverse report from BOB.[67]

When it reached the floor in mid-June 1938, the bill was revised at the request of several Democratic senators to further emphasize that the major responsibility for promoting travel was vested in Interior (presumably reflecting Ickes's behind-the-scenes lobbying). Despite BOB's opposition, the bill passed without debate and by unanimous consent on June 13.[68] It was a flat-out victory for Ickes. But then the bill died in the House because the Interstate and Foreign Commerce Committee took no action on it before the adjournment of the Seventy-Fifth Congress three days later.[69]

4

When Business Liked (Part of) the New Deal, 1939

Business support for USTB became increasingly apparent when, early in the year, Ickes invited industry representatives to a meeting to help boost the bureau and urge passage of enabling legislation. Attending and expressing support for an active federal involvement in promoting tourism included the American Merchant Marine Institute, TWA airline, American Hotel Association, Association of American Railroads, Chrysler and Ford car companies, American Automobile Association, US Chamber of Commerce, and American Express.[1] For the industry, it was a revelation that government could be on its side and help it make more money. For New Dealers, it was a fantasy come true of finally being able to reach across the political abyss considering the virulent opposition from the private sector for so much of what the administration sought to accomplish.

Throughout 1939 (and particularly after Hitler's invasion of Poland on September 1), the bureau noted the impacts of the international situation as good news for domestic and hemispheric travel. Yes, fewer European tourists were likely to come to the United States, but that also meant fewer Americans would travel abroad. They could replace their plans with, for example, domestic travel, trips to Puerto Rico (no passports needed), and Caribbean cruises. Also, USTB suggested, "as a result of the war in Europe," trips to Latin America were increasing in popularity and would be patriotic, as that would help implement FDR's Good Neighbor policy.[2] Indicating the close relationship between travel promotion and public relations, the federal government's delegation to a conference on inter-American travel included, beside a USTB representative, the PR director of the US Maritime Commission, Robert Horton (Lee 2012, 20). In general, the bureau expected it would be a "wonder year" for domestic tourism, with its offices reporting that business was brisk.[3] USTB's chief predicted that, "with Europe now roped off to tourist travel," trips within the United States in 1939 would likely top 1928, the last full year of prosperity before the Great Depression.[4]

Only toward the end of the year did the wartime situation abroad begin to cast a pall on this exuberance. The San Francisco world's fair closed early, on October 29 instead of the planned closing on December 2. Furthermore, the plans that the fair would reopen in 1940 were suddenly "indefinite."[5] The European and Asian conflicts also prompted the federal government in December 1939 to increase security around key sites, such as the Boulder (Hoover) Dam. USTB issued a press release emphasizing that these new security restrictions did not ban visitor access to the dam, which was a popular tourist destination, along with other major engineering marvels of the day.[6] To counter a looming tourism slump, USTB released a statement from First Lady Eleanor Roosevelt. She acknowledged that "travel in other countries is not possible at the present time and may not be for some time to come. Therefore, let us utilize our vacation periods to really see our own country." And no matter where you went in the United States, everybody spoke English![7] In fact, Miami reported a substantial increase in visits and hotel bookings that it "traced almost directly to the European war."[8]

In November, in an attempt to mobilize additional external support for USTB legislation, Ickes addressed the annual conference of the American Automobile Association (AAA). He called on the organization to endorse legislation creating USTB because so much of domestic tourism was by car.[9] But he shot himself in the foot politically when, almost gratuitously, he suggested that AAA also seek legislation or regulations to prohibit trucking on weekends and holidays so that tourists and travelers would not be competing with these "road hogs."[10] Wisconsin's Public Service Commission, which regulated trucking as well as utilities, indeed had such a policy, and it considered the five-day trucking week a success. Naturally, Ickes's comment prompted severe criticism from the trucking industry and this dominated the media coverage.[11] His speech was broadcast on the Mutual Broadcasting national radio network, so his call for public support for USTB legislation could have had an impact on public opinion. But the side controversy diluted the effect of his main message. He had committed the PR mistake of stepping on his own message. Unrepentant, Ickes said in his diary that the speech "went over very well" even though "it aroused a great deal of controversy."[12]

Organization and Leadership

In March, Ickes appointed Ruth Bryan Rohde as USTB's collaborator for one dollar a year as the successor to Gerard (who had resigned in mid-1938 [chap. 3]).[13] Rohde, a former two-term member of Congress (D-FL) was the daughter of William Jennings Bryan, the Democratic nominee for president in 1896, 1900, and 1908, as well as President Wilson's first secretary of state. After her defeat for reelection in 1932, FDR appointed her minister to Denmark, the first woman to hold this diplomatic rank. Upon returning to the United States, she was in touch with Mrs. Roosevelt who, in turn, suggested to Ickes that Rohde could help pass the travel bill due to her standing with non–New Deal "old line Democrats."[14] This group thought well of her father, Rohde implied, even though they generally opposed the expansion of the federal government under FDR. Perhaps she would have the credibility to convince them to go along with such legislation. Ickes, fervently committed to passing a bill to give Interior the travel portfolio, was game if it might help. However, Rohde was disappointed at the unremunerated appointment. She had hoped for a paid position, such as for a lecture tour of South America to encourage visits to the United States.[15] Nonetheless, she accepted the position.[16]

Her initial duty was to go to San Francisco as part of an official State Department delegation to a hemispheric conference on travel, coinciding with the upcoming San Francisco fair.[17] On her way there and back, she gave talks and interviews in Los Angeles and Chicago promoting travel.[18] She also wrote a column promoting USTB for the *Christian Science Monitor*.[19]

But, a few months later, Ickes peremptorily fired her. It was part of a major shake-up he imposed on USTB's organizational status and leadership in the spring of 1939. In short order, he moved the bureau to be under his direct supervision, demoted Loomis, named a new head of the bureau, and then fired Rohde and Loomis. First, in late March, Ickes transferred USTB from the National Park Service to the Office of the Secretary (Boyd 1941, 253). Given how strongly he wanted his department to have the travel portfolio assigned to it through legislation, Ickes feared keeping the bureau in NPS would weaken his case. For example, businesses involved in tourism would want legislation to promote all travel, not just to national parks. This would be a particular concern for state tourism bureaus and local chambers of commerce that did not have any major NPS parks or historic sites nearby. A comment by a member of the powerful House appropriations subcommittee controlling Interior Department funding reflected this perspective. At a hearing in February, Congressman Albert Carter (R-CA) said about USTB that "there are a lot of fine things that are done these days that I would not advocate be done by the National Park Service" (US House 1939a, 711). Ickes was preemptively trying to help get the bill passed by removing the bureau from NPS. (There may well have been other reasons, such as his level of confidence in the leadership of NPS to do a good job overseeing the bureau and/or Park Service displeasure with this responsibility.)

Second, around the same time, he demoted Loomis to be only the chief of the DC office, and he appointed W. Bruce Macnamee, then a special assistant to the secretary, as bureau director.[20] In his diary, Ickes said Loomis "has been highly unsatisfactory." Amusingly, Ickes felt the same way about Macnamee. He "doesn't seem to be very industrious," but "I think that he ought to be able to do this job [USTB] with greater satisfaction to himself and profit to the department."[21] A few months after that, Ickes suddenly fired Rohde and Loomis. Ickes claimed that his appointment of Rohde had been contaminated by a political recommendation from Mrs. Roosevelt to hire her, thus giving the scent of partisan patronage to the role instead of apolitical merit. At the same time, he fired Loomis, claiming that Loomis had improperly tried to influence legislation by urging private groups to lobby Congress for the travel bill. This rationale was odd, to say the least, given how much Ickes wanted the legislation to pass. Perhaps Ickes feared that Loomis's grassroots lobbying efforts would be exposed and would be fatal to the legislation. Congress didn't like federal agencies beating the grassroots bushes to lobby Congress in favor of an agency.[22] There was also an insinuation that Loomis had engineered Rohde's appointment and therefore had to bear the consequences of it (yet it was Ickes who had announced it). Loomis counterclaimed that he had resigned already, that he did not understand why he was being fired, that his work had been praised, and furthermore that a USTB staffer who was a relative of NPS director Demaray, had written derisively of the agency's work.[23]

It was a big confusing mess, with one paper neutrally calling it a "travel row."[24] The great differences in headlines of the same United Press wire story gave a hint at how perplexing and convoluted the story was:

- "Ickes Fires Official of Travel Agency"[25]
- "First Lady's Plea of Mrs. Rohde Vetoed, Ickes Travel Row Reveals"[26]
- "'Lobbying' Is Attacked"[27]
- "Job for Ruth Bryan Rohde to Influence Old Line Democrats Urged by F. D."[28]
- "Ickes' Plan for Tourist Trade Boost Hit by Row in Ranks"[29]

A few days later, the anti-FDR *Chicago Tribune* printed its own negative version of the story: "Mrs. Roosevelt Gets Ruth Bryan Federal Position."[30] But, in all, it was a confusing set of developments including charges, countercharges, and denials. Perhaps because it was such a complicated narrative, the story as a political scandal died out quickly. The speed with which Rohde's firing came (and went) led to some reporters not being aware that she was no longer with USTB—and she did not correct them, either.[31]

There were hints that these upheavals were causing organizational problems, noticeable in the capital. In July, a nationally syndicated columnist wrote that Loomis was the fall guy in those events and that his departure was a "bureaucratic tragedy."[32] The next month, another *Post* columnist (whose beat was the federal civil service) enigmatically described the bureau as "embattled," an apparent reference to the changes that Ickes had imposed a few months earlier.[33] Toward the end of the year, Macnamee had to send an apology to an NPS information officer about the bureau releasing incorrect information on accommodations at national parks. He blamed "the personnel situation in the Bureau" for the problem and said that, for the future, "undoubtedly it would be wise" for USTB staff to be in touch with her directly whenever they received public inquiries about NPS rates and accommodations.[34]

Notwithstanding these disruptions, USTB's staff and budget expanded and solidified in 1939. It received $70,000 of already appropriated funds to NPS to promote visits to the parks (US House 1939c, 22) and partly through WPA's allocation of personnel funded under the 1933 Emergency Relief Act (ERA) (US House 1939a, 703). Indicating significant growth, the federal agency in charge of providing office space in the District of Columbia reported that USTB's Washington office had sixty employees and occupied 5,300 square feet (US House 1939b, 9). In mid-1939, FDR approved WPA grants of $35,000 to the bureau to prepare and disseminate additional materials, especially for increased visitors to San Francisco for the world's fair.[35] The bureau's partnership with WPA also expanded that year with USTB promoting the publication of state-by-state guides written and published by the WPA Federal Writers' Project.[36] (For further discussion of the USTB-WPA partnership, see chap. 5.)

The bureau also began preparations for an expensive glossy color brochure on tourist attractions in the United States with an estimated budget of $200,000. The publication was intended for distribution abroad and would consist of sections on destinations in each region of the country, with equal space for each region. In a widespread mailing to its stakeholders, the bureau invited participation in submitting material and photos for use in the brochure, as well as cost-sharing contributions (USTB 1939a).

Public Relations

The focal points for USTB's promotional themes in 1939 were the New York and San Francisco world's fairs, encouraging foreign tourism to the United States, and—particularly due to European tensions—urging citizens to travel domestically, to American territories (administered by Interior) and to Latin America. The last issue of USTB's *Official Bulletin* in 1938 kicked off these 1939 themes by presenting FDR's formal proclamation inviting citizens and foreigners to visit the fairs.[37] Throughout the year, the bureau used the slogan "See America First" as a unifying theme for domestic tourism and an alternative to foreign travel.

Throughout the year, USTB tried to maximize its efforts to promote travel by using as many platforms and media as possible. It issued a brochure that invited the public to contact it for information, describing itself as "a travel service" (USTB 1939b). At midyear, it inaugurated a semi-annual *Calendar of Events* (USTB 1939c), with regular updated and expanded supplements, as well as a directory of *State Travel Information Agencies* (USTB 1939d). The bureau also launched biweekly newsletters from its two field offices. The San Francisco office began distributing a newsletter oriented to western audiences called *Travel News*. Its debut issue was on June 1. The New York office's *Travel and Recreation News Letter* tailored its content for eastern readers. Its inaugural issue was on August 19.[38] Parallel to those regional publications, USTB continued publishing its national *Official Bulletin* on a monthly basis until July (nos. 4–10). In issue 11, it increased frequency to twice a month (usually the 10th and 25th) and changed the format "for a larger and somewhat more readable type." It also made permanent a new front-page format, featuring a guest column (and portrait photo) written by a leader in one of the major travel-related associations and organizations. The purpose was to bring to "our readers discussion of travel problems by those who are most familiar with them."[39] This feature had the benefit of endearing USTB to its private-sector stakeholders who saw their particular line of work promoted and in their own words. These initial biweekly issues continued to be four pages, but the last issue of 1939 (#20) had twelve pages, indicating growth of content, interest, circulation, and federal commitment.

Continuing and expanding its PR practices from 1937, the bureau issued press releases,[40] obtained coverage in specialized publications,[41] provided feature information for the Sunday papers,[42] released new research data on travel,[43] targeted backpacking as a growth sector,[44] and authored columns distributed to the press as features.[45] It also worked with Interior's Office of Education to encourage school tours.[46] Indicating the growth in the public's utilization of the Bureau's services, the New York office reported that it had handled about 18% more requests for information in 1939 than in 1938. To reflect USTB's effort to be integrated with the for-profit travel sector and state tourism offices, over the year it had distributed a quarter of a million maps and brochures provided by those sources. It also distributed 508 posters and loaned 142 movies.[47] Regarding print material, illustrated brochures, booklets, and pamphlets were especially popular.[48]

African Americans

During his first full year at the New York office, McDowell focused on promoting travel by African Americans through information dissemination. He hired two African American women to staff his office. Anna German was his personal secretary, and Rosalind Boston was a receptionist/clerk.[49] To organize his work (and receive the implied formal and official imprimatur of the government), he prepared and released a statement on "The United States Travel Bureau and Its Relation to Negro Travel." He also circulated a "General Plan for Travel Clubs." The latter called for creation of more travel clubs oriented to African Americans and campaigned for "better hotels and guest houses, clean and attractive restaurants," and other facilities necessary for African Americans to enjoy travel.[50]

McDowell's most important publication in 1939 was a *Directory of Negro Hotels and Guest Houses in the United States, 1939* (USTB 1939e). The twelve-page directory listed the establishments that welcomed African American travelers, presumably most being segregated. It listed formal hotels and private homes that rented rooms in thirty-nine of the forty-eight states, as well as all black branches of the YMCA and the YWCA. This was a very spare publication, in typescript and mimeographed. It listed no place of publication, was not formally published by the US GPO, and had neither a preface nor an introduction. The only text, other than the listings themselves, was the disclaimer that the bureau did not "imply a recommendation" for the establishments it listed. Nonetheless, the publication of the *Directory* "showed an unprecedented sensitivity to the needs of African Americans" (Armstead 2005, 155). The New York office reported distributing 131 copies of the publication in 1939, presumably proactive mailings to mailing lists prepared by McDowell as well as reactive responses to individual requests.[51]

McDowell occasionally gave talks out of town, issued press releases, and was active in business associations—all circumscribed to African Americans. The world's fair in New York was a major travel attraction for African Americans as well as for whites. For example, McDowell met a group of African American high school students from the Washington DC area at Penn Station in Manhattan and accompanied them during a weekend visit to the fair.[52] For a Negro National Business conference to be held in New York to coincide with the Fair, McDowell served as vice chair of the planning committee.[53] McDowell gave a talk in Greenville (SC) to highlight the release of "a guide for colored motorists" who wanted to visit the New York world's fair. It had been prepared by the city's Colored Professional and Business Men's Club and was titled *Dixie to the New York World's Fair*. In his talk, he said that "there was an increased interest among white travel agencies in the promotion of group Negro tours" and pledged USTB's support for such initiatives.[54] In November, he issued a press release that a new hotel for African Americans had just opened near Savanah (GA). It was owned and operated by the former maid to the late actress Marie Dressler, using funds willed by the actress for this express purpose, evidently to assure her a livelihood for the rest of her life.[55]

McDowell sometimes wrote columns, usually titled "Travel Notes," for the newsletter published by USTB's New York office. He was identified as being with the bureau's "Division of Negro Affairs." His columns were placed near the back of the issues. These details all hinted at the political touchiness of the subject of race. By limiting his column to a regional publication of eastern readers rather than the national *Official Bulletin*, and by burying his column near the back of each issue, the bureau was trying not to attract political attention and attacks. Judged by the norms of racial segregation of the times, the bureau's slightly hidden support for promoting African American tourism deserves commendation nonetheless.

In McDowell's first column in August, he reported that as a result of his efforts, nineteen more travel agencies catering to African Americans were being opened, including seven in New York. He hoped that these efforts would lead to "fuller enjoyment" of travel by African Americans.[56] In an October column, he noted that Joe Louis, heavyweight boxing champion, had sponsored a horse show in Chicago, and he wrote of a dude ranch in California that welcomed African Americans (presumably operated by an African American).[57] The next month, he informed readers of a new Florida resort, opened by the Afro-American Life Insurance Company of Florida, and another in New Jersey, operated by a "colored business woman" from Philadelphia. He also praised the record-setting attendance at the Texas State Fair on "Negro Day," celebrating the fifty-first anniversary of African American participation in the fair.[58] Later columns in 1939 praised a five-hundred-member travel club organized at a night high school in Washington DC and a facility in Hot Springs (AR) that was "the only institution of its kind owned and operated by Negroes . . . and is one of the best Colored hotels in the United States."[59]

Toward a Statutory Travel Bureau: Seventy-Sixth Congress, First Session, 1939

For Ickes, the adjournment of the Seventy-Fifth Congress in 1938 must have felt like a heartbreaking near miss. The Senate had passed a bill giving USTB statutory status and placing it in Interior. But the House Commerce Committee did not act on it. With a new Congress, now it was back to square one. Ickes was determined to get it through in the 1939–40 Congress. As soon as the new Congress was sworn in, he arranged for identical bills to be introduced in the Senate by Josiah Bailey (D-NC) and again in the House by Clarence Lea (D-CA).[60] A magazine for the automobile industry promptly applauded the good news.[61]

Ickes quickly convened an ad hoc group of executives of the travel industry. As mentioned at the beginning of the chapter, attendees included representatives of the American Hotel Association, American Express,

American Steamship & Tourists Agent Association, Association of American Railroads, Automobile Manufacturing Association, National Association of Motor Bus Operators, US Chamber of Commerce, Air Transport Association, and American Automobile Association.[62] It was an impressive indication of the scope of the political powerhouse that the industry could exert in support of a New Deal tourism promotion program. At the meeting, they informally endorsed the bills.[63] Here was a concrete indication of how much business had come to support this particular FDR program. Setting their conservative ideology aside, if it was good for them, it was a good idea—FDR or not, New Deal or not.

USTB's newsletter also trumpeted the endorsement of the legislation by smaller newspapers outside the New York–Washington axis, many of which were conservative editorially and FDR critics. They included the *Fresno [CA] Bee* and the *Winston-Salem [NC] Journal*.[64] The travel reporter for the *Times* wrote that "according to all the legislative signs," the bill would be passed during this Congress.[65] A columnist for the *Post* thought so, too. He noticed the "curious" lack of controversy or opposition to the bill. The subject generated an unusual political alliance, noting that Senator Bailey was an "ultra-conservative" and that the bill seemed to have "no kick" (i.e., opposition) from the major influential lobbies on Capitol Hill.[66] A North Carolina newspaper also noticed the odd couple of home-state Senator Bailey with the liberal Ickes. They were "together engaged" for the same goal.[67] However, media support was not unanimous. For example, the conservative *Manchester [NH] Union* still did not see the need for creating USTB.[68] A nationally syndicated conservative columnist mocked Ickes's sudden love of business, which previously he had routinely denounced on other issues, such as when he criticized utility companies as inherently opposed to the public interest. That the USTB bill was endorsed by business was "a peculiar argument for the hater of 'economic royalists' who demand too much return on their investments, in his opinion."[69]

The House Commerce Committee held a hearing on the bill in late March. Supporting testimony from the private sector came from the American Hotel Association and the Air Transport Association, as well as endorsements from federal agencies, including the Civil Aeronautics Authority, US Maritime Commission, and the State Department.[70] President Roosevelt also wrote a letter endorsing the general purposes of the bill, although, cagily, not explicitly committing himself to any particulars (US House 1939c, 10). The Commerce Department broke the seeming unanimity of support. As in years before, it insisted that any tourism promotion should be housed in its Bureau of Foreign and Domestic Commerce. It argued that travel promotion was a broader mission than merely visiting national parks. But it had a weak hand, stating in the hearing that even if the bill were to be passed giving jurisdiction to Commerce, the Department could still not engage in any such activities without new funding from Congress. A congressman argued back, saying that it "sounds to me like a case of the grocer selling groceries, and you wanting to give the butcher the right to say how the grocer shall advertise" (US House 1939c, 9). Interior seemed to have not only the high ground politically, but also financially. It argued that NPS's current budget was already providing some funding for USTB (to promote visiting its parks), and therefore the importance of new funding was somewhat less pressing than Commerce's.

A few weeks later, in April, the House committee endorsed a slightly modified version of the bill, reflecting some of the comments at the hearing, including authorizing $100,000 a year to USTB.[71] By now, in its legislative clearance role, the Bureau of the Budget removed one roadblock, stating that the bill was consonant with the administration's legislative program. Though short of a full endorsement that encouraged passage, this removed a red or yellow light from the president's office (US House 1939d, 3).

Before the bill could reach the House floor for debate, a subcommittee of the Senate Commerce Committee held a hearing on the companion bill.[72] At the time, the bill was viewed as having been thoroughly vetted by committee approval in the preceding Congress. Therefore, it held only a brief token hearing. However, the Commerce Department maintained its opposition to the bill, terming it "a radical departure" from the longstanding precedents that commerce-related activities would be assigned to the Commerce Department (US Senate 1939, 8).[73] Disregarding that objection, at the end of the hearing, the subcommittee recommended that the full committee approve the bill. Its draft for the committee's formal report endorsed the pending House version and called the legislation "a modest, constructive effort to increase travel within the United States." It also rebutted Commerce's objections, saying that Interior was "the logical place to put it" because of the close link of travel to promoting visits to NPS venues. Therefore, "in the interest of economy and good administration it is deemed best to have these two agencies under one administrative head" (12).

While Ickes may have won that round, Commerce's opposition was consistent and could kill the bill down the line. In the multistep legislative process, opponents only have to win once, while proponents have to win at every stage. A senator made the point bluntly in letters to the two departments. Unless they "got together on this bill, it would not be approved."[74] This made explicit how politicians viewed the world. They much preferred win-win situations than binary zero-sum choices. In the latter situations, one side would be happy with them, and the other, equally unhappy. There was no political benefit to being forced to make the choice. Politics preferably is aggregative, about building political credit and not spending it for unnecessary reasons.[75] Better to force a compromise than to take sides.

Ickes's near obsession to pass the bill had to overcome this roadblock. However, beside the bureaucratic turf warfare, there were some related hurdles. In December 1938, FDR had appointed Harry Hopkins as secretary of commerce. It was an odd appointment, given that Hopkins up until then had only served in social welfare positions and had occasionally engaged in New Deal antibusiness rhetoric. The Senate vote to confirm him was 58–27, indicating how unpopular he was with the congressional conservative coalition. Although Ickes and Hopkins were both dedicated to the New Deal and loyal to FDR, they were political adversaries and rivals for primacy in the administration. They often competed with each other, such as the seemingly overlapping missions of Ickes's PWA and Hopkins's WPA. A pointed example of this occurred in mid-1939, when FDR appointed former Indiana governor Paul McNutt to head the new Federal Security Agency. Reflecting his suspiciousness of Hopkins, Ickes grumbled in his diary about Hopkins inviting McNutt to his Maryland farm for a weekend stay. Ickes was suspicious that Hopkins was wooing McNutt to affiliate himself with Hopkins's political network within the administration (Ickes 1974, 686). Ickes was afraid that Hopkins might not want to yield regarding the travel portfolio simply due to their political rivalry.

Compounding the turf warfare and competitiveness within the administration, there was a further problem for Ickes if he was to overcome Commerce's opposition to the USTB bill. Hopkins was very sick at the time. While the illness was never definitively diagnosed, he was homebound at his Maryland farm for months and sometimes so weak he was bedridden. Ickes desperately wanted to talk to Hopkins about the jurisdictional dispute, but phone calls to Hopkins's office were futile because he was "never at his office," and when he tried to reach him with long-distance calls to Hopkins's farm, he could not get him to come to the phone. Ickes's frustration reflected his persistence and own workaholism. In mid-1939, he twice complained in his diary about how hard it was to reach Hopkins (Ickes 1974, 682, 687). Ickes was hounding Hopkins, practically chasing him around Washington.

Perhaps Hopkins realized that the only way to get Ickes off his back was to deal with the issue.

Ickes's single-mindedness and harassment of Hopkins finally paid off. In June, he somehow got Hopkins to sign a letter stating that "while the present bill places the function of promoting travel in the Department of the Interior, after carefully considering the importance of the legislation and its value to the industries concerned, this Department favors the enactment of the proposed legislation."[76] For Ickes, it was a home run and stood out as one of the few changes in Commerce's departmental policy that Hopkins adopted while secretary. In part, it may have been that Hopkins simply did not care about this particular policy issue nor the department forsaking its claim to being the primary agency to promote travel. In general, Hopkins was not particularly interested in the minutia of public administration, whether with WPA (and its predecessor, the Federal Emergency Relief Administration) and, later, Lend-Lease (Lee 2017b, 30–31; 2018, 132). He was more of a policy guy than a bureaucratic empire builder. It is also possible that he did not have the strength to fight with Ickes about it and perhaps thought a political favor to Ickes might get paid back in the future regarding a policy matter that Hopkins cared more about. Perhaps he just wanted Ickes to go away and leave him in peace.[77]

Nonetheless, Ickes's hope to fast track the bill failed. It did not come to the floor until late July. By then, the House Commerce Committee issued another report endorsing a further modified version of the original bill (US House 1939e).[78] On July 31, 1939, the House debated the bill. Notwithstanding some critical comments by conservatives, no one was willing to oppose the bill flatly. Even for conservatives, their states wanted to benefit from increased tourism. In part, the floor manager credited the lack of overt opposition to the "carpenter work" done by the committee to remove or reduce as many concerns and objections as possible. After a short debate, it passed by unanimous consent.[79]

Then, nothing. The Senate committee did not act on the House-passed version of the bill, even though it had already held a perfunctory hearing in April, and the subcommittee had recommended adoption, including having submitted a draft for the full committee's report (US Senate 1939). Macnamee and Ickes were still optimistic the bill would come out of committee and get a floor vote in the fall session,[80] but there was no action on the bill in committee, let alone on the floor.[81] An internal USTB document blamed the inaction on an "impasse within the [Senate Commerce] Committee," including an "indifference" to the subject. It is, of course, possible that the inaction may have been because of a much more important subject deserving Senate attention, the outbreak of World War II in Europe on September 1, 1939, when Hitler invaded Poland (and the Soviet Union following up a few weeks later). The internal USTB memo tried to turn that perspective upside down with the parochial argument that the war "has given tremendous impetus to the travel-America idea."[82] (For legislative developments in 1940, see chapter 5.)

5

Congress Decides It Sometimes Likes Agency PR

Statutory Creation, 1940

For a nation not at war, 1940 was a tumultuous and eventful year, both abroad and at home. Germany conquered Denmark and Norway in April and then invaded Belgium, Holland, and France in May. The withdrawal of troops from Dunkirk occurred in late May and early June. France surrendered on June 22. The Battle of Britain began in the air in July and lasted until nearly the end of the year. These events meant, of course, that tourism from Europe to the United States was down by 95 percent.[1] It also meant tourism from the United States to popular destinations such as France and Britain also plummeted to near zero. Through it all, USTB tried to adapt its goals to reflect changing and new realities.

USTB adjusted to these fluid circumstances by promoting the slogan of 1940 as "Travel America Year." In mid-January, President Roosevelt signed a proclamation noting that "the exigencies of international conflict may be expected to deter travel by American citizens to the areas involved." On the other hand, "no such deterrent to travel exists among the friendly nations of the Western Hemisphere." Therefore, he hoped that "we in the Americas further consolidate our unity by a better knowledge of our own and each others' countries through the instrumentality of travel."[2] In part, this was a diplomatic effort to implement his earlier "Good Neighbor" policy by strengthening relations between Latin America and the United States.[3] His carefully worded slogan of "Travel America" was deliberate. It could be interpreted as encompassing travel to and from all the Americas, not just domestic tourism within the United States. As a result, Macnamee was actively involved in forming a hemispheric travel organization that included the United States, Canada, and Latin American countries.[4]

Increases in domestic tourism would also serve the earlier goals of the administration's travel policy, including expanding visits to national parks and other federal landmarks, thus promoting support for FDR's conservation policies. Similarly, increasing revenues to the travel business could promote more business support for the New Deal. Headlines such as "U.S. Travel: War Only Changes the Direction," "Hotel Leaders Praise 'Travel America Year,'" and "Tour Heads Back 'Travel America'" all captured the economics and politics of the situation.[5] Even business bastion *Barron's* trilled at the commercial profits to be reaped domestically due to the war overseas.[6]

Other factors were at play. Wartime—even if the United States was not a combatant nation—inevitably led to an increase in patriotism and love of country. What could better serve and enhance this public mood than domestic travel? Also, the war abroad was having a significant positive economic effect at home leading, indirectly, to expanded travel. Great Britain was buying large amounts of materiel and food in the United States (for cash, as this was before Lend-Lease) and the president's declaration of a limited national emergency (in late 1939) was the beginning of huge increases in army and navy spending on rearmament. As a result, unemployment dropped, incomes increased, and consumers had more discretionary income to spend on vacation travel.[7] Finally, politics was also lurking in the background. There would be a presidential election in November 1940, with intense partisanship in Washington and political positioning (including the issue of whether or not FDR would run for a third term). Economic prosperity, increasing patriotism, and support for the administration's travel promotion by private interests would inevitably have political and electoral payoffs for FDR and Democrats.

In all, 1940 was a great year for travel and tourism. As early as July, Macnamee predicted that it would be a good year, notwithstanding the international situation. He acknowledged the national mood in the initial months of 1940 by saying that "people were jittery at first over the war. Now they're saying: 'Let's go out and have fun while we can.'"[8] By October, visits to national parks were setting new records,[9] and railroads were particularly benefiting from increased travel.[10] At the end of the year, USTB said it had been the "greatest year in American travel history."[11]

Organization and Leadership

In 1940, the bureau published a revised and updated version of the previous year's introductory brochure (USTB 1939b). It adopted a revised motto for public use and a kind of mission statement for internal decision making and planning. The new slogan was

Strengthen America: Travel Promotes the Nation's
- Health
- Wealth
- Unity (USTB 1940a, 3)

Indicating its *quid pro quo* relationship with the travel industry, the brochure was at pains to explain what services USTB provided and what it did not. "The Travel Bureau does not furnish transportation rates, book reservations on tours and cruises, or suggest specific highway routes or eating and overnight accommodations. Such services are personally and efficiently handled by the numerous steamship and tourist agents, automobile clubs, railroad, bus and air line ticket agents" (USTB 1940a, 3).

For most of 1940, USTB had to rely on funding and personnel that largely came from WPA's emergency relief appropriations, with "a few positions" funded by CCC (US House 1940a, 712). Nonetheless, it had been able to expand into a modest-sized bureaucracy, with eighty-four full-time staffers. Macnamee's headquarters office in DC had twenty-six employees (not counting him); thirty worked in the New York office, and twenty-seven in San Francisco.[12] Its budget for FY 1940 (i.e., July 1939–June 1940) was $108,000, with about three-fourths coming from WPA and the rest from CCC. For FY 1941 (July 1940–June 1941), the bureau's budget was $112,000, with the funding from WPA and CCC in about the same proportion (US House 1941a, 557).

The bureau's location in the Interior Department's organization changed again in 1940 because Ickes was being buffeted by conflicting political currents. It will be recalled that in March 1939, he had transferred it from NPS to the Office of the Secretary. There, it would operate as a freestanding entity within the department. Reflecting that, the bureau's three newsletters did not mention any affiliation with NPS. This related to the travel industry wanting clarity that USTB promoted all domestic travel, not just to national parks.

Another reason might have been at play: a new international mission for USTB. At midyear, FDR asked Congress to approve supplemental funding to strengthen federal efforts at closer relations with Latin America. His request included $25,000 for USTB to promote inter-American tourism (US Senate 1940a, 2). The purpose was to cover the costs of translating and rescoring existing federal films and publications on US tourist attractions into Spanish and Portuguese and then arranging for wide distribution in Central and South America. Congress approved an appropriation of $12,500.[13] This expansion of the bureau's work to include promoting international travel from Latin America to the United States was wholly unrelated to USTB's original assignment of promoting domestic travel only. Impliedly at least, this also meant USTB was to promote travel by US citizens *to* Canada and Latin America, another new mission wholly unrelated to domestic tourism and/or national parks.[14] Therefore, Ickes may have felt that shifting USTB's organizational home to the secretary's office made more sense than keeping it in NPS.

But Ickes's near-fanatical desire to get Congress to pass a bill creating a statutory USTB in his department led to counterpressures. During the year, he quietly moved the bureau back to NPS. This was reflected in most of the bureau's publications, which identified it as a unit within NPS. Why? It was a political expedient to get the bill through Congress. Conservatives who opposed the expansion of federal activities during the New Deal said that they would vote against any bill that created a new agency within the executive branch. Therefore, as a matter of principle, they would oppose the USTB bill (US House 1941a, 555). There was a neat bureaucratic legerdemain and bill-drafting trick to solve that problem. By reinserting USTB into the National Park Service, and stating this explicitly in the legislation, the bureau was now—presto!—not a new agency; it was merely a new program activity within an existing agency.

Public Relations

It will be recalled that USTB was largely financed by WPA allotments of personnel positions and small cash grants for specific purposes. During 1940, the bureau became a major consumer of some of WPA's noninfrastructure work products: travel posters and state guidebooks. Here was a powerful synergy between two separate New Deal programs, both helping the other be successful, both helping promote economic growth.

Perhaps the most-remembered PR campaign by USTB was its poster series. Using WPA artists, these posters provided strong visual signals of the physical attractions of a particular state. The project began in early 1940, when the bureau asked WPA's Poster Division (located in New York City) to assign some of its artists to design posters for a "See America" series. One of the earliest was a pictorial map of the United States with graphics identifying the kinds of activities that travelers could find in various regions, such as hiking, fishing, sightseeing, scenery, music festivals, and sunbathing. WPA considered that poster so creative and important that it reproduced the map on the back cover of a handbook helping its artists use the silk-screen printing process (Velonis 1940, back cover). Another of the initial posters designed for USTB in 1940 was "See America: Visit the National Parks" (USTB 1940b, 15).

Richard Floethe, director of the Federal Art Project's Poster Division, considered USTB such a major client that in a list of almost a dozen governmental agencies that sought its design services, he listed the bureau second, after the US Public Health Service (Velonis 1940, Introduction [n.p.]). The "See America" poster series began with about six posters and the pictorial map, all with an underlying theme that "exemplifies the depression era search for a usable past that could ameliorate social tensions and unite Americans by recovering and affirming national values" (Pillen 2008, 49). (See figures 1–7 in the color gallery following page 62.) One indication of the long-lasting impact and signal quality of these WPA-USTB travel posters occurred in 2017. That year, the US Postal Service released ten commemorative stamps of the best WPA posters. One was USTB's "See America: Welcome to Montana" poster, displaying a rider on horseback admiring the outdoor scenery of forests and snow-capped mountains. The poster conveyed a "wilderness aesthetic. The man, alone in his reverie and immersed in the undeveloped landscape, embodies the freedom and sense of isolation that many associated with a 'primitive' wilderness experience, one distinguished from the hustle and bustle of city life" (Pillen 2020, chap. 3).

Later in 1940, USTB involved itself further in the travel-poster business by creating a central clearinghouse and catalog for all travel posters circulated by state travel offices, localities, railroads, steamship companies, and all other sources.[15] For example, in a three-paragraph announcement in *Women's Wear Daily*, the bureau said it had free posters available for use by retailers in their street-level window displays on recreation and travel wear.[16] Within two weeks, it got 100 requests from stores.[17] By the end of the year, USTB listed 213 posters it had available in its clearinghouse for distribution (USTB 1940b).

The Federal Writers' Project (FWP) was another noninfrastructure WPA program, in this case to hire writers. Like public works, construction, and conservation projects, it had a dual goal: providing jobs to the unemployed and having them create tangible work products that benefited society. In this case, one of FWP's major initiatives was the production of the American Guide Series, state-by-state volumes summarizing all major aspects of a state. While not the sole purpose of the guides, each state volume could be used as a guidebook for travelers and tourists, providing interesting information about the state, whether notable sites, history, or other important information. USTB jumped at the opportunity to promote the series as useful, low-cost, nonparochial, government-sponsored guidebooks. Beginning in 1939, the *Official Bulletin*'s book review section often promoted recently published volumes of the state series.[18] By 1940, nearly every issue of the bureau's newsletter from January to September reviewed and plugged the latest state guides that had been recently released.[19] To increase public use of the guidebooks, the bureau maintained a complete run of the state series in its San Francisco office's library.[20] Griswold also noted that the close relationship between FWP and USTB included a FWP author writing radio scripts for USTB's use in promoting the benefits of the American Guides (2016, 83).

FWP also published other books outside the state series that were useful to travelers, including guides to a city and guides to a specific historical site. USTB also promoted those books. For example, in 1939, FWP released and USTB's *Official Bulletin* reviewed such volumes as *Skiing in the East: Ski Trails and How to Get There* and *New York City: A Metropolitan Playground*.[21] These cross-agency marketing promotions continued in 1940, with the book-review section promoting volumes such as *Cape Cod Pilot* and *Mission San Xavier Del Bac, Arizona*.[22] Twenty years later, an academic examination of FWP's American Guide series concluded that the individual volumes were of varying quality but, as a whole, the series should "merit considerable respect" including serving as a "well-drawn road map" for travelers (Fox 1961, 19).

The bureau's PR activities activities in 1940 expanded on its 1939 practices, including press releases, radio broadcasts, exhibits, and information desks at the two world's fairs (both of which did reopen for second seasons in 1940), a lending library of free travel-related movies, newsletters, literature, event calendars, and information booths and exhibits at travel shows. The volume of requests for information increased in 1940, with the bureau receiving and responding to 146,000 inquiries (US House 1941a, 557).

Macnamee continued speaking to conferences of the travel industry[23] and submitting guest columns to the travel sections of the weekend newspapers.[24]

The bureau continued publishing new reports of original research and statistics, presumably of intense interest to its external stakeholders and the travel reporters for newspapers. One report focused on the effect of recreational travel on land use (Dorsett and Johnson 1940) and another on the volume of inter-American tourism (Dorsett 1940). Another bureau staffer wrote an article about railroad publications as outlets for travel-related public relations (Donavan 1940). At the end of the year, the bureau committed to publishing research reports on trends in travel because of "fast-developing and ever-changing" patterns that those in the business needed to be aware of. Publishing such research data would "permit some knowledge of probable changes in the future" and consequently facilitate adaptive behavior by the bureau's stakeholders.[25] Indicating the usefulness of this research as well as its expanding base of stakeholders, a reporter from *Women's Wear Daily* wrote for its annual special issue on the upcoming summer season about the implications of some of the bureau's research results. He highlighted some likely trends by travelers and, consequently, their clothing and accessory needs.[26]

Some expansions of PR activities in 1940 included the New York office opening a larger and more permanent "Travel America Hall" that conveyed the "drama, romance, and value of touring America" (USTB 1940a). Probably quoting a USTB press release, one paper reported that it was the agency's "the first major activity" since its establishment.[27] Other federal agencies wanted to get in on the travel action. Interior's Grazing Service (!) sent an exhibit to display in the New York office's travel hall, and the Civil Aeronautics Authority (CAA) provided an exhibit on its role in promoting more air travel at the San Francisco world's fair.[28] The San Francisco office moved to larger quarters that included an assembly hall and lecture room for screening travelogues and other public events, as well as a library of travel materials and research reports.[29] An example of a new PR focus was to popularize snow skiing. Both regional offices released new brochures to promote the sport. The seventeen-page directory from the New York office listed private skiing locations as well as those at major national parks (USTB 1940c). The publication from the San Francisco office was more ambitious, with forty-two pages of descriptive information about western venues, as well as a calendar of events related to the sport (USTB 1940d).

In at least one case, USTB public relations shrank in 1940. For reasons that are unclear, *Travel and Recreation News Letter*, prepared by the New

York office, suspended publication in March,[30] while its counterpart from the San Francisco office, *Travel News*, continued throughout 1940.

African Americans

Beginning in 1936, William Green, an African American who lived in New York, edited and published annual editions of *The Negro Motorist Green Book*. He was "an entrepreneurial and eloquent Harlem-based letter carrier" who initially collected much of his information from African American letter carriers in other cities.[31] The book contained lists of establishments around the country that welcomed African Americans. The title page of the 1940 edition displayed something new: "Prepared in cooperation with The United States Travel Bureau" (Green 1940). (See figure 8 in the color gallery following page 62.) By utilizing the materials McDowell had collected for USTB's 1939 *Directory*, the 1940 *Green Book* had new and more extensive listings.[32] The bureau's national periodical, the *Official Bulletin*, contained a brief article noting the publication of the *Green Book* and the contribution the bureau made to it.[33] It was placed on the last page of the issue, the second-to-last item, and was one of the very few references to African Americans in USTB's national newsletter.

The 1940 *Green Book* also contained a three-page article written by McDowell—identified as USTB collaborator—on his travels from New York to Washington, Greenville, Atlanta, and Savannah (McDowell 1940). He made part of the trip by Greyhound bus and part on the Central and Georgia Railroad. (Presumably, after crossing the Mason-Dixon Line, he had to sit in segregated locations.) McDowell identified some places of interest to African Americans, including "many beautiful homes, owned by Negroes" in Greenville, Decatur Street in Atlanta, the "large number of Negro W.P.A. workers" at an archeological dig in Irene Mound, "one of the oldest Negro churches in North America" in Savannah, and the story of the African American captain of a ferry to sea islands off the South Carolina coast. Here was "a Colored man, exercising much authority . . . a sight seldom seen elsewhere." His travelogue was an effort to identify interesting sites for African American tourists and to demonstrate that such traveling could be accomplished without major racial problems.

During 1940, McDowell continued writing a "Travel Notes" column for African Americans in the *Travel and Recreation News Letter* published by the New York office. In a January column, he provided statistics on the amount of money spent by African American tourists, estimating annual expenditures of $140 million.³⁴ His columns in the two February issues included a long quote by famed scholar W. E. B. DuBois about his travels in Florida, followed by a description of a California resort for African Americans in the Los Angeles area.³⁵ McDowell also worked with the New York World's Fair to have a "Negro Day" on August 15, 1940.³⁶ During the spring, he served on a local planning committee for it and expected about fifty thousand African Americans to visit the fair that day.³⁷ The events of Negro Day included a national beauty contest called Miss Sepia-America. McDowell assured African Americans that USTB "will go to great lengths to assure the general public their utmost cooperation," presumably a reference to otherwise predictable white hostility and discourtesy when large numbers of African Americans congregated in public venues.³⁸

A Statutory Travel Bureau: Seventy-Sixth Congress, Third Session, 1940

USTB's image on Capitol Hill continued being positive, probably reflecting its grassroots support in the travel industry. For example, a senator inserted into the *Congressional Record* a statement praising the new travel initiative to promote inter-American tourism and a congressman reprinted the text of a national radio broadcast complimenting the national parks, including USTB's role in promoting visits to them.³⁹

Conservatives, however, viewed USTB as part and parcel of the evils of the New Deal, particularly FDR's expansion of the federal bureaucracy and his creation of scores of new alphabet agencies. For example, in a floor speech, a Republican congressman denounced the establishment of seventy new agencies during the Roosevelt administration, thus creating "the world's greatest bureaucracy."⁴⁰ USTB was on his list. However, by now, Ickes had moved USTB back into NPS, and therefore, arguably, it was not a new freestanding agency. But for this conservative, such a detail would have been a distinction without difference and therefore not persuasive enough to get USTB off his list. The congressman's general critique of these new agencies was "their seizure of control of the Nation's industrial, commercial, and educational activities."⁴¹ This broad-brush criticism must be seen as somewhat accurate. USTB was quietly co-opting the private sector's travel business as well as trying to coordinate tourism promotion efforts by state and local governments. For example, the bureau described its work as "a *centralized* partnership between business and government."[42] The congressman was actually not far off the mark because USTB was an example of a new federal role and agency where none had existed before.

Similarly, beginning in the 1890s, Congress became increasingly concerned about public relations by executive-branch agencies. Agencies, like Victorian children, should be seen and not heard. Good deeds would speak for themselves. And if anyone needed to be a link between the public at large and the federal government, well, that was what congressmen and senators were there for. The negative view of government propaganda became increasingly sharp during the New Deal. FDR had expanded not only the role of government but also the voice of government. Each new agency tended to have a well-staffed information office, bent on telling the citizenry about its programs and accomplishments. The vitriol from the congressional conservative coalition against federal PR became intensely pointed and denunciatory (Lee 2011). Yet proposals for federal promotion of domestic travel seemed to be an exception to the Congress's anti-PR rule. No, this was *not* propaganda! This was a service to the citizenry and the businesses depending on them. This was good PR and needed to be enhanced. Taking it one step further, congressional authorization of a federal office to promote tourism would be for an agency whose *raison d'être* was to do PR. External communication was its only product line. This did not seem to raise the usual red flags on Capitol Hill. Presumably, the business constituency lobbying for the bill had the effect of muting the nearly reflexive congressional criticism of federal PR.

It will be recalled that in mid-1939, the House of Representatives had passed a bill establishing a statutory USTB in the Department of Interior without significant debate and by unanimous consent (chap. 4). But there had been no action on it in the Senate. One of the problems was that Senator Josiah Bailey (D-NC), the ostensible champion of the bill (he had introduced an earlier version of it in January 1939) was chair of the Senate Commerce Committee, and apparently, he was not using his chairpersonship to get the committee to take action on it. Behind the scenes, in February, a committee staffer reported that there was a consensus by members that "the time is not ripe" for the bill, presumably reflecting the chair's perspective.[43] From the outside trying to look in, Interior was flummoxed. Ickes speculated to the president of Bailey's "indifference" to the bill,[44] while NPS officials heard rumors that Bailey's inaction was due to a reason that was "personal," that is, political.[45] Unknown to Ickes, Bailey told a committee staffer that he was "not in favor" of the bill, even though it was an iteration of the bill *he* had introduced in 1939.[46]

FDR passed Ickes's letter on to BOB. Two weeks later, he signed a draft prepared by BOB's legislative clearance office.[47] He said BOB had some modest requests for revisions in the bill. If these amendments were added to the current version, then the president had "no objection" to the bill nor to Ickes publicly using the arguments for such legislation as he had stated in his letter.[48] The president's tepid and passive acquiescence to a modified bill was all that Ickes needed. And Ickes was nothing if not persistent. He *wanted* this victory.

Meanwhile, a middle-ranking member of the committee, Senator Theodore Bilbo (D-MS) had—also tepidly—asked Interior for more information about the bill. Bilbo, a crude racist,[49] was not viewed as a legislative workhorse, but nonetheless sometimes involved himself in moving legislation that was unrelated to his parochial state's (and racial) interests. He was also chair of a subcommittee of the Commerce Committee, with the prosaic and catch-all title of Subcommittee on Fisheries, Forest Products, Minerals, and Land Surveys. (It did not have jurisdiction over the USTB bill.) Bilbo's low-key inquiry triggered a quick and detailed response from Macnamee. He thanked the senator for what Macnamee characterized as Bilbo's "evaluation of the importance of the Bureau" and provided a summary of the benefits of the bill. Probably impoliticly, Macnamee said, "it seems incredible that this legislation might be allowed to linger in Committee because of inertia or a misconception of its purposes."[50] For reasons that are unclear, in late 1939 and early 1940, several committee reports recommending legislation to the Senate were signed by Bilbo, not Bailey. That might be explained because many came from his subcommittee. But that was not the case for the USTB bill. Nonetheless, on March 20, Bilbo was the signer of the formal report on the USTB bill to the Senate. He stated that the committee had acted on the bill, recommended some amendments, and "as so amended, recommend that the bill do pass" (US Senate 1940b, 1).

The likely explanation is that Bailey no longer supported such legislation and did not want the responsibility of promoting it on the floor of the Senate. The committee report did not contain a minority report stating objections of any committee members to the bill. But the vote was not unanimous. While there is no record of the committee session and vote, minority party member Arthur Vandenberg (R-MI) later said he had opposed to the bill in committee and that majority party member Bennett 'Champ'

Clark (D-MO) similarly had "formidable objections" to the bill and had expressed them "at some length" during the committee's executive session on it.[51] According to a committee staffer, the action on the bill occurred only because some committee members "brought it up" during a meeting of the full committee and hijacked the chairman's agenda. Bailey had not scheduled the bill for executive action, but he was unable to prevent it from happening when the committee was meeting to act on other bills. The staffer also said that the recommendation to pass the bill reflected the preference of only a slight majority of the members there, just enough that "there were a sufficient number of them to carry the motion to report the bill," further implying it was contentious.[52]

Ickes was nonetheless elated. He quickly pivoted by lobbying Senate Majority Leader Alben Barkley (D-KY) for a prompt scheduling of the bill for floor consideration.[53] Ickes depicted the bill as noncontroversial, with no opposition from other government agencies and broad support from the private sector. But when Barkley tried to call up the bill on April 10 by asking to do so by unanimous consent, Vandenberg objected.[54] It was the parliamentary equivalent of a veto. Without unanimous consent the bill could not be debated at that time. It would have to get in line, a long and slow moving one, particularly given that the presidential nominating conventions were just a few months away and partisanship was especially high.[55] Not one to fold in the face of opposition, Ickes called Vandenberg. The senator explained that his opposition was a general and principled one against new spending and new agencies. But he promised Ickes that "he would not fight us very hard." Having moved USTB back to NPS, Ickes could counter that the bureau would not be a new agency, and therefore Vanderberg's opposition would not apply in this case. Having resolved that (and ignoring Vandenberg's claim on the floor that Senator Clark also opposed it), Ickes now asserted that there was "no other opposition" to the bill in the Senate.[56]

Therefore, two months after that failure on the floor, Ickes asked Barkley to try again.[57] On June 22, Barkley asked unanimous consent to bring up the bill. No one objected. Ickes's lobbying had worked. There was a brief debate. Barkley made a point of emphasizing that USTB would be an entity within NPS (not a new federal agency) and that the bill only authorized USTB to have a budget of up to $100,000 a year, but did not appropriate it any funds. That decision would be up to the Appropriations Committee. Oddly, he claimed that the bill had been recommended unanimously by the committee. Bilbo also spoke in favor, saying that "this is the best we can do at this juncture to give business not only to the hotels in the United States, which are all behind the proposal, but to the public generally. It is a great scheme, and the small amount asked is infinitesimal when compared to what some other governments spend." The bill was adopted by voice vote.[58] As there were minor differences between the versions passed by both houses, a conference committee was necessary. It quickly ironed out the differences and recommended passage of a revised version (US House 1940b). Both houses did so.[59] Without ceremony or public statement, FDR signed the bill on July 19, 1940.[60]

Understandably, the new law received no attention from the daily press. It was overshadowed by a flurry of bills deemed much more important that he signed over the course of a few days. Regarding the war, he signed into law a $40 million fund for war risk insurance for merchant shipping and $4 billion for a two-ocean navy. News coverage of politically oriented bill signings focused on the expansion of the Hatch Act (prohibiting political activities by civil servants) to state and local government employees funded by federal grants. But even those important bill signings were overshadowed by external news: the beginning of the Battle of Britain and FDR announcing on July 16 that he would accept nomination for a third term. Minor coverage of the USTB bill signing occurred in the news magazine for the railroad industry and, two months later, in the Sunday travel section of the *Herald Tribune*.[61]

USTB had made it across the congressional finish line. This was no mean feat, and the saga reflected the difficulties inherent in the law-making process as well as Ickes's never-say-die persistence. The venue for promoting travel was now officially sanctioned as belonging in the Interior Department, not in Commerce. Thus, promotional activities would originate from within NPS, not from a new agency.[62] With USTB as a statutory entity, it would now be eligible for congressional appropriations during the normal annual funding process, rather than relying on indirect funding from emergency spending bills for relief, employment, and conservation. USTB had become an ordinary and routine federal program and entity.

Figure 1. Welcome to Montana [horse rider and mountains], See America campaign, USTB poster and stamp from WPA Poster series, 2017. Poster credit: Martin Weitzman, New York City: Federal Art Project, WPA, about 1940. Stamp credit: United States Postal Service®, © United States Postal Service. All rights reserved.

Figure 2. Welcome to Montana [horse rider & teepees], See America campaign, USTB poster. Credit: Richard Halls, New York City: Federal Art Project, WPA, about 1940. Accessed January 10, 2019, http://www.loc.gov/pictures/item/96503139/

Figure 3. Welcome to Montana [teepees & trees], See America campaign, USTB poster. Credit: Jerome Rothstein [aka Roth], New York City: Federal Art Project, WPA, about 1940. Accessed, January 10, 2019, http://www.loc.gov/pictures/item/98518516/

Figure 4. Visit the National Parks, See America campaign, USTB poster. Credit: Harry Herzog, New York City: Federal Art Project, WPA, about 1940. Accessed January 10, 2019, http://www.loc.gov/pictures/item/98518589

Figure 5. [Arches National Park], See America campaign, USTB poster. Credit: Frank S. Nicholson, New York City: Federal Art Project, WPA, about 1940. Accessed January 10, 2019, http://www.loc.gov/pictures/item/93505613/

Figure 6. [Carlsbad Caverns National Park], See America campaign, USTB poster. Credit: Alexander Dux, New York City: Federal Art Project, WPA, about 1940. Accessed January 10, 2019, http://www.loc.gov/pictures/item/96503125/

Figure 7. [Pictorial Map of US], See America campaign, USTB poster. Credit: Aida McKenzie, New York City: Federal Art Project, WPA, about 1940. Courtesy of The Wheatley Press.

Figure 8. Cover, *Negro Motorist Green-Book*, 1940. Credit: Digital Collections, New York Public Library, Schomburg Center for Research in Black Culture, Jean Blackwell Hutson Research and Reference Division. No use restrictions. Accessed January 9, 2019, https://digitalcollections.nypl.org/items/dc858e50-83d3-0132-2266-58d385a7b928

Figure 9. Cover, *Negro Motorist Green-Book*, 1941. Credit: Digital Collections, New York Public Library, Schomburg Center for Research in Black Culture, Jean Blackwell Hutson Research and Reference Division. No use restrictions. Accessed January 9, 2019, https://digitalcollections.nypl.org/items/cc8306a0-83c4-0132-cc93-58d385a7bbd0

6

Promoting Tourism during a National Emergency, 1941

If 1940 had been a tumultuous year in wars abroad (and politics domestically), 1941 was even more so.[1] Germany conquered Yugoslavia, Greece, and Crete beginning in April 1941. Rommel's forces in North Africa pushed east and imposed a siege on Tobruk. Hitler's invasion of the Soviet Union began on June 22, and, by the beginning of winter, his armies conquered astounding amounts of territory and captured or killed millions of Soviet soldiers. Japan's efforts to conquer more of China continued in 1941, with it holding most of the eastern coastal areas. Japan also extended its military control over (Vichy-ruled) Vietnam by invading southern Indochina in July.

In response, FDR asked Congress to approve enormous amounts of money for arms production and military mobilization. Also, in March, he succeeded in convincing Congress to enact Lend-Lease, which was essential for Great Britain to be able to continue fighting. As soon as Russia was attacked, he expanded Lend-Lease to provide equipment to the USSR. In August, the one-year limit on the time that draftees could be kept in the service was about to go into effect. By a one-vote margin, he persuaded Congress to extend it indefinitely. To oversee these domestic national defense programs, he appointed a new White House official to oversee the Office for Emergency Management as a major coordinating agency within the Executive Office of the President (Lee 2018). FDR also strengthened control of arms production and economic priorities by creating the Office of Production Management and the Office of Price Administration. To promote national morale and domestic preparedness, he established the Office of Civilian Defense. These were defensive and preparedness activities that maintained some degree of normalcy in day-to-day life. Also, the United States continued to have diplomatic relations with Germany and Japan, and was not in a state of war against either of them.

However, in 1941, the United States became increasingly a de facto combatant in the European war, engaging in military activities that were more than passive or defensive and that extended beyond the Western Hemisphere. Throughout 1940 and early 1941, the country had been in a limited national emergency.[2] Then, in spring 1941, FDR declared that the United States was in a state of *unlimited* national emergency (Roosevelt 1969, 1941: 194–95). For example, in April, the president announced the US takeover of Danish Greenland, and, in July, American armed forces occupied Iceland. He declared large swaths of the Atlantic as under US naval protection and extended US convoy escorts as far east as Iceland for merchant ships delivering military and civilian supplies to the United Kingdom. The USS *Greer* ineffectively engaged a German submarine in early September. On September 11, he ordered the navy to shoot on sight any German or Italian vessels, even if they were not at that moment attacking any American ship. In October, German subs torpedoed and sank two navy vessels.

These events created a new environment for daily life in the United States in 1941. Through it all, the travel business in general and USTB specifically tried to promote continued and even expanded tourism and travel. They clung to an increasingly out-of-touch illusion of normalcy. What, specifically, was the proper role of leisure travel in a time of national emergency, whether a limited one (before May 27) or, now, an unlimited one?

Convincing Congress Again, this Time for Money: Seventy-Seventh Congress, 1941

Getting the USTB law signed in 1940 was only half the battle. It had merely *authorized* funding for a new statutory entity. Now came the hard part: getting money appropriated to the bureau. It was almost as though there were parallel congressional universes: one for law-making, the other for funding. The power of the purse was exercised with extreme scrutiny by the House and Senate Appropriations Committees. As required by the Constitution, all funding bills had to start in the House, giving its appropriations committee first crack at funding bills. Then, bringing up the rear, was the Senate. As a result of this required sequence, the Senate Committee often acted as an appeals court for agencies unhappy with decisions by the House. That was exactly what happened with USTB's first interaction with the appropriation process for FY 1942 (scheduled to begin on July 1, 1941).

The other built-in advantage that the House Appropriations Committee had over its Senate counterpart was size. Given that the House had 435 members versus 96 senators, each senator had to cover more policy bandwidth and accept more committee assignments than representatives. That meant senators were stretched more thinly. For example, in the Seventy-Seventh Congress, each house member had roughly one to three (major) committee assignments, while a senator served on about four to six committees. In turn, the House Appropriations Committee during the 1941 session had 42 members, while its Senate counterpart had only 23. Therefore, as a result of the Constitutional sequence of appropriations bills and sizes of committees, individual House members could delve into the arcane details of agency operations more deeply and with more impact than senators.

Generally, the annual funding of the federal government was accomplished through a baker's dozen of appropriations bills, some covering a single department, some covering a passel of them. The Interior Department had its own annual appropriation bill, and it was overseen by House and Senate Appropriations Subcommittees focusing solely on it. The members of these subcommittees were a kind of budget bureau for the legislative branch, and they could not have cared less about a new agency having been authorized by their parent institutions. That did not obligate them. They would pass a *de novo* judgment about this new (public administration) kid on the block and make their own decisions if it should be funded at all. Then, if they decided it should be funded, they could decide by how much. Authorization was only the ceiling on the how much could be appropriated. Any amount from zero to the authorization maximum was fair game for the subcommittee appropriators.

The annual budgeting process routinely began in January, when the president would submit his budget proposal to Congress for the fiscal year to begin six months hence. On January 3, 1941, FDR sent Congress his FY 1942 spending plan. In it, he proposed that the new USTB be appropriated $75,000 for the year (US House 1941b, A57, A105) and have approved staffing of twenty-seven people, nine in the DC headquarters and eighteen at the field offices (US House 1941b, 600–01). Regarding personnel, the background budget justification documents stated that its staffers currently funded by WPA and CCC would be rehired by the bureau for its permanent staff (US House 1941a, 555).

The House Appropriations Committee's Subcommittee on the Interior Department held a hearing on NPS's budget request in March. Testimony focused on the fact that USTB distributed "disinterested" information in (quiet) contradistinction to materials from the travel industry which were parochial (556). Charles Leavy (D-WA) asked NPS associate director Demaray how USTB would handle a request from him about driving to Los Angeles? Demaray replied that for such a specific query, "They probably would advise you to secure accurate road information from the A.A.A. [American Automobile Association]." As a general policy and following the protocols with the travel industry, USTB would not recommend if a traveler should get there by train, plane, bus, or car. His answer was intended to clarify that USTB did not compete with nor duplicate nongovernmental sources of information. But what he said made it sound like USTB was little more than a postal forwarding station and therefore of limited benefit to individual members of the public. Leavy pressed Macnamee to identify more clearly what actual service USTB provided directly that would benefit "the average American citizen" (i.e., voter)? Macnamee muffed his answer, emphasizing that USTB worked in a cooperative basis with a $6 billion industry. So, parried Leavy, it sounded like the bureau mostly served "those people who are engaged in the field as a commercial enterprise." Macnamee lamely interjected that the bureau had answered about 150,000 queries the preceding year. But his answer did not explicitly state that this statistic was of queries that were completely fulfilled by USTB versus those that merely confirmed receiving it and that the request was being forwarded to a nonprofit association or for-profit business to handle (556–57).

In another damaging exchange, Robert Jones (R-OH) asked, "You advertise this service over the radio?" Macnamee answered, "Yes." In that case, "Have you any money provided for that purpose?" "We have not so far," said Macnamee (557). Jones's question could have been interpreted as relating to *paying* for air time to advertise (begging the question why a government agency should even do that) or only the cost of *preparing* public service (i.e., free) advertising. Macnamee's replies sounded like USTB had long-range plans for major radio advertising expenses. The implication was that the funding request for FY 1942 was only the beginning of a federal agency that would be expanding significantly in the future and seeing greatly increased appropriations.

When Secretary Ickes testified before the subcommittee on the departmental budget request as a whole, Albert Carter (R-CA) asked him about USTB's plans for increasing foreign visits to the United States. For example, regarding Latin America, "Do you have agents there?" No, said Ickes, there was no funding for it. That's why USTB cooperated with existing travel agencies (presumably referring to for-profit businesses). But, in the long-run, he hoped to establish USTB offices in major foreign countries to promote visiting the United States and handling general queries comparable to the tourist offices that many European and other counties had in the United States (or had had before the war) (1115).

In all, these brief and seemingly fact-based exchanges were inauspicious for the bureau. The consequences emerged when the House Appropriations Committee submitted to the full House its recommendations for Interior's FY 1942 budget. For USTB, it recommended no funding at all:

> It is the opinion of the committee that any necessary data in this connection is available from private travel bureaus, hotels, chambers of commerce, automobile associations, and other agencies of a like character, and that the undertaking of this activity as a permanent obligation on the part of the Federal Government is unjustified. (US House, 1941c, 18)

Ickes was no doubt disappointed. It was as though he had to run the same hurdle course he had run to get authorizing legislation, except ending with the opposite result. The House decision attracted little media attention,[3] although an editorial in the *Times* encouraged the Senate to restore the funds because the bureau "seems to have done a good job" and "has won the confidence of those commercially interested in travel."[4]

But he had another crack at it, by appealing to the Senate's subcommittee for Interior appropriations to reinsert the money and (hopefully) in a conference committee prevail on the House to recede from its position (US Senate 1941a, 13). For this round, Ickes was in luck. The chair of the subcommittee was Carl Hayden (D-AZ), from a state with a disproportionately large tourism-based economy. He would be interested in protecting his state's interest in promoting travel.[5] Hayden showed his hand at the beginning of the hearing on the USTB budget. He said that he had "received a number of letters and telegrams, the tenor of which is that there has been some such service sometime rendered and taken away" (216). Indicating the flowering of business support for USTB, Hayden invited representatives from the American Hotel Association and the National Bus Traffic Association who had come to the hearing to testify for USTB (219–22). Then he inserted into the hearing record eight more supportive letters and telegrams, including two from Arizona (223–26). There was a moment of wobbling when Pat McCarran (D-NV) asked how much in total was spent by governments, nonprofits, and private businesses to promote domestic travel. Macnamee said he didn't have that figure (226–27). However, Nevada also had a substantial tourism-based economy, so Macnamee's lack of knowledge was not fatal.

In early June, the Senate Appropriations Committee recommended that the Senate restore in full the president's request for $75,000 for USTB (US Senate 1941b, 9). In conference committee, there was no pressure to split the difference. Instead, the House receded, agreeing to the Senate's position of $75,000 (US House 1941d, 9). President Roosevelt signed it into law in late June, just days before the beginning of the fiscal year.[6]

There were a few lesser budget skirmishes with Congress in 1941 relating to USTB's role in promoting tourism from Latin America. It will be recalled that in 1940, FDR had requested $25,000 for USTB to rescore federal films into Spanish and Portuguese and then give them wide distribution in South America. Congress had approved half that (covering three new staffers). Now the president's budget request for FY 1942 included continuing the program with full funding of $25,000 for another year and hiring a fourth staffer (US House 1941e, 318). The House Appropriations Committee's subcommittee for the State Department minutely scrutinized this, including examining Macnamee and NPS's Cammerer in excruciating detail (US House 1941e, 389–97).[7] It eventually recommended $20,000 (US House 1941f, 13). Again, the Senate Committee heard an appeal to restore the $5,000. The discussion of this small amount comprised three pages of the hearing record (US Senate 1941c, 66–68). It recommended restoring the $5,000 cut and the conference committee accepted the Senate position. FDR signed the bill into law in late June.[8] Also, as part of an omnibus deficiency bill, Congress approved a request from the State Department for $3,500 for USTB to lead a seven-person delegation to the Second Inter-American Travel Congress in Mexico City in the fall of 1941.[9]

Organization and Leadership

Administratively, the bureau was a relatively stable operation in 1941. Its organization chart was unchanged from the previous year, with a headquarters office in DC and regional offices in New York and San Francisco. Macnamee continued as bureau chief. The heads of the field offices had the title of supervisor. Early in the year, Jay Wingate was named to head the New York office, and Horton S. Allen was appointed its assistant supervisor.[10] J. L. Bossemeyer continued as supervisor of the San Francisco office. Based on the definitions it used, the US Civil Service Commission considered Macnamee, Wingate, and Bossemeyer as holding senior administrative and executive positions.[11]

For FY 1942, USTB's annual operating budget totaled $103,500: $75,000 for routine operations, $25,000 for Latin American outreach, and $3,500 for participation in a hemispheric travel conference. Its staffing based on those appropriations was twenty-seven people for regular operations and an additional four for the South American program. As of the fall of 1941, ten of them were based in the DC office (US House 1944, 380). WPA continued providing some assistance to USTB during the first half of 1941, including the distribution of leaflets and pamphlets from the headquarters, as well as staff to help the New York and San Francisco offices publish their newsletters.[12] This support ended at the conclusion of FY 1941 on June 30. To cover for this loss of WPA funding, NPS transferred $2,880 from its reserve funds to USTB to hire two junior clerk-typists in the Washington office to handle the general distribution of leaflets and pamphlets.[13]

Public Relations

Based on the activities USTB had developed in previous years, in 1941, it continued engaging in and expanding a catholic range of public relations and promotional activities. For example, USTB's audio-visual activities expanded. The New York office regularly screened film travelogues every weekday from late morning to early afternoon.[14] It estimated weekly attendance for the films was about 2,500 to 3,000 people.[15] The New York office also offered free loans from its movie collection to groups interested in travel.[16] There were also occasional screenings in Washington.[17] The bureau produced a forty-five-minute travelogue of Texas, which a film magazine called "an outstanding travel film for all age groups."[18]

Beside movies, USTB sought to be a central source of other visual materials on tourism and travel. The San Francisco office accumulated a library of five thousand color slides available for loans, display, and publication.[19] From its library of still photos, USTB provided pictures for a college textbook (published the following year). Indicating the scope of that collection, some of the photos were credited directly to USTB, and others were stock photos in the bureau's library from the Louisiana Tourist Bureau, Virginia Chamber of Commerce, New Mexico Tourist Bureau, and the Santa Fe Railroad (Barnes and Ruedi 1942, 25, 34, 169, 170, 217, 340, 342–43, 534).

The bureau's radio profile was enhanced in midyear when it was featured for an installment of the nationally distributed federal radio series *U.S. Government Reports* (US Congress 1942, 1154).[20] Other PR activities included distributing free maps and posters and making available several traveling exhibits.[21] Trying to take advantage of any and every PR opportunity, when two movie theaters in New York City were conducting an "I am an American" promotion, the regional office placed displays in their lobbies of scenic locations in the United States worth traveling to.[22]

To promote winter skiing, in advance of the 1941–42 season, the San Francisco office published an elaborate thirty-five-page directory of skiing locations including inserts (USTB 1941a). Also, the New York office sponsored a "ski service"[23] and regular five-minute radio updates on skiing conditions in the East every Friday.[24] Ambitiously, the bureau encouraged citizen expectations of going on two vacations a year. Besides the traditional summer vacation, winter vacations—for such activities as skiing—could now become the norm for American families.[25]

Macnamee continued traveling, giving speeches promoting tourism,[26] and releasing newspaper columns.[27] The bureau also kept up issuing a stream of press releases with upbeat news about travel in 1941. These were often covered as spot news and included in the Sunday travel sections.[28] In June, the beginning of the summer travel season, USTB released a directory of travel destinations, mostly on the East Coast. According to the *New York Times*, it was "a long list of places where holiday-seekers may make merry."[29] Acknowledging receiving such friendly and continuing media coverage, in mid-1941 Ickes said, "The press and radio have been generous in their consistent support of the Bureau's current travel promotion program" (Ickes 1941, 298).

USTB's publication program included a semi-annual calendar of events (USTB 1941b), supplemented by monthly updates.[30] In June, it released a list of summer events.[31] The *Official Bulletin* continued, although somewhat erratically. At the beginning of the year, it was a monthly publication; in March, it shifted to bimonthly; it ceased after the July–August issue; then it resumed as a monthly in November. Most issues were twelve pages, the last two issues were sixteen and twenty pages respectively. It will be recalled that in March 1940, the New York office had ceased publishing its *Travel and Recreation News Letter*. In March 1941, the publication resumed with the rechristened title *Eastern Travel Today*. Originally intended to publish two issues a month,[32] it never reached that goal. It was published on a monthly basis, but then it suspended publication from July to September. It resumed in October.[33]

On the other hand, San Francisco's biweekly continued uninterrupted in 1941. In March, its title changed from *Travel News* to *Travel West*. The name change was part of an upgrading of the professional look of the publication, shifting to a more polished look with a photo as the front cover. Most issues were twenty to thirty pages. The name change also reflected an increasing self-consciousness by western states that the powerful and mythic image of the American West in pop culture could be invoked for travel promotion. The theme of the campaign developed by business was "See the Old West This Year." USTB strongly supported this initiative. Bossemeyer, head of the San Francisco office, gave a talk at an industry conference to organize the effort. He encouraged each state and locality to focus on the distinct and unique sights each could promote as destinations.[34]

At the urging of the travel industry, eager to increase business, USTB participated in an effort to promote congressional approval of a law moving some federal holidays to Mondays so that travelers could have several three-day weekends during the year.[35] This PR campaign reflected a kind of business-as-usual effort by the bureau in 1941. But as the year unfolded with increasingly serious national and international developments, the goal gradually came to look trivial and even selfish. The travel editor of a newspaper, normally a cheerleader for the industry and USTB, criticized the idea as "more than debatable" and the arguments for it as "entertaining" if irrelevant.[36]

USTB's broad efforts to link to, and promote, all aspects of the travel business were exemplified in its outreach to the luggage industry. The *Official Bulletin* in the spring featured a full-page article written by the head of the industry's professional association. Glad to benefit from the bureau's PR, he noted the profits from what he called "World War number two," in that Americans shoppers no longer could buy imported European suitcases and now were giving their business to American manufacturers.[37] A few months later, in observance of National Luggage and Leather Goods Week (!), the New York office changed its street-level window display to give prominence to luggage loaned to it by a nearby suitcase store.[38]

An indication of external stakeholder support for, and satisfaction with, USTB's PR came from one of its enthusiastic backers. The president of the American Automobile Association (AAA) complimented the federal government's new role of playing "a more active part" in promoting travel to the public (Henry 1941, 903). Given that so much of travel in the United States was by privately-owned cars, AAA's good opinion of the bureau can be interpreted as a third-party and independent confirmation that USTB's PR was important and effective to the economic beneficiaries of its work.

African Americans

By 1941, the mission of promoting domestic travel by African Americans had become a relatively routine activity by the New York field office, managed by McDowell as the collaborator of the Division of Negro Activities. One of the activities he continued in 1941 was preparing an updated version 1939's travel directory for African Americans. The *Negro Hotels and Guest Houses, 1941* (USTB 1941c) was slightly improved, including more content (eighteen versus twelve pages) and a more professional-looking typeset layout. It also had a welcoming preface that the 1939 version did not have. The preface stated that the goal of the publication was to be "a reliable source of information for Negro travelers," acknowledging that USTB's promotion of travel by African Americans was closely tied to "the extent that only the assurance of adequate accommodations can effect." Similar to the 1939 edition, besides listing hotels, the directory included Negro Branches of YMCAs and YWCAs.

Another recurring activity was assisting in the issuance of an updated edition of the *Green Book*. Like the 1940 edition, the title page credited the book as "Prepared in cooperation with The United States Travel Bureau" (Green 1941).[39] (See figure 9 in the color gallery following page 62.) The political delicacy of an administration agency promoting the interests of African Americans was reflected in an article in the *Official Bulletin* about the publication of this new edition. The article was very short (two paragraphs), was placed on the last page of the issue, and was the last article on that page. It appeared to be aimed at white readers and accepted the status quo of segregation: "Everybody who sees the Negro motorist bound on long trips wonders how they find accommodations that serve them. The answer is this guide."[40] There was no consternation about segregation nor encouragement to hotels to stop being restricted to whites. Also, the article did not acknowledge that the bureau had played any role in helping prepare the new edition.

McDowell continued writing for the New York office's restarted and renamed publication. As in 1940, his columns were placed near the end of each issue. In March, he wrote about some state parks in the South that were dedicated to use by African Americans. For example, he noted an Arkansas state park that was "developed as a recreational area of Negroes" and that South Carolina had set aside some land "as a Negro State Park." He had developed a comprehensive list of such state facilities and invited readers to write him for copies.[41] His April column announced an upcoming conference of the Bureau of Colored Work of the National Recreation Association and invited anyone who worked in recreation to come. The conference was planning to focus on recreational needs of African Americans now in the military and those working in defense industries. The goal was to assure them that there were "recreation programs and opportunities available to colored groups."[42] In his next column, McDowell highlighted "Negro Day" at the annual "Lost Colony" summer pageant at a federal park in North Carolina. He also noted that it was near "the only Coast Guard station in the United States manned by Negroes." As part of the events that day, they would present a demonstration of life-saving skills.[43]

Instead of featuring a column by McDowell, the May issue presented a guest column by James A. Jackson, a special representative of Esso marketers. (Presumably, he was African American, and his assignment was to promote patronage of Esso gas stations by African Americans.) Jackson doubted that "the American public is aware of the extent to which Negroes travel by motor." He suggested more African Americans would travel "were it not for handicaps encountered by this group in connection with some existing touring facilities."[44] McDowell also continued visiting cities with a substantial African American population and, when there, dropping by the offices of the local African American newspapers. During the summer, he visited the offices of Baltimore's *Afro-American* to promote USTB's new guide for state parks open to African Americans.[45]

A hint of McDowell's fate came when, after its summer hiatus, *Eastern Travel Today* resumed publication in the fall. The October and November issues did not have his usual column. Apparently, he was no longer on the staff of the New York office. However, it wasn't until early December 1941 when there was confirmation. An article distributed from Washington by Associated Negro Press (ANP) reported that USTB had ceased its employment of McDowell.[46] It provided no explanation for the change. (Similarly, the agency's records at the National Archives did not have any documentation on this.)

It is unlikely that the reason was budget cutbacks. In the late fall of 1941, USTB was in the midst of FY 1942, and its appropriation had already been approved by Congress for the fiscal year that began in July. After that, Congress had not imposed any specific budget cuts on civilian agencies to counterbalance the enormous new sums it was approving for national defense. So the reason for McDowell's release would not have been due to legislatively imposed budget cuts. Similarly, the reason could not have been wartime retrenchment because the article was first published on December 1, a week before Pearl Harbor and the US declaration of war on Japan. It is unclear from the article if the reporter had talked to McDowell. If he did, it is possible that McDowell may have felt it was wiser to stay quiet than publicly criticize USTB. Being diplomatic would keep his options open for the future, whether to be rehired by the bureau or to demonstrate to any other future employer his trustworthiness and loyalty. The ANP story praised McDowell's accomplishments, that he "generally serviced the colored traveling public" and "did quite a job," including the "well prepared" hotel guide.[47] Thus ended USTB's targeted outreach to African Americans to promote domestic tourism.[48] Notwithstanding the bureau's pioneering efforts, issues of legal and social restrictions on African American tourism continued after WWII (Armstead 2005, 155–57) and through to the civil rights era.

National Defense Gradually Crimps Recreational Travel

All its PR activities and coalition-building notwithstanding, the topic impinging on USTB in 1941 more and more was national defense. At first, the bureau and the industry in general thought that the European and Asian wars could be exploited as a benefit. If Americans could not travel to Europe or Asia, perhaps they could be convinced to tour parts of the United States or take American ships on cruises to the Caribbean or Latin America.[49] Similarly, Macnamee urged East Coast Americans who routinely traveled to spas in Europe to redirect their trips to equally high-quality spas in the United States, such as Saratoga (NY) and Hot Springs (AR).[50] This all sounded like it could be good news, if one was willing to disregard the significance and implications of what was happening in the real world.

The bureau had started 1941 optimistically with the slogan of "Travel Strengthens America."[51] In January, Macnamee wrote a column titled "Travel Looks Ahead to Another Big Year."[52] Similarly, the head of the San Francisco office wrote that 1941 had the "prospects for a great travel year."[53] A guest column in the bureau's *Official Bulletin* declared that travel was integral to "the American way of life."⁵⁴ Impliedly, it was un-American not to travel, even in these times. At a travel conference in February, USTB fine-tuned its current message. Now travel was even more important because "recreational travel [is] at a time when every one should be as fit as possible [and] is an important part of personal defense." In general, continued travel was important because it "builds health, wealth and unity."⁵⁵

By spring, reality was beginning to encroach on all this happy news. In a March column, the head of the San Francisco office defensively wrote that travel was important economically, and "it is even more important to the social and political welfare of the people." Reductions in travel would not only have a direct impact on the economics of the industry, but also the "national morale would be adversely affected if our people were deprived of the stimulus they derive from travel, and from the rest and recreation that it would afford to individuals during these trying times." Nonetheless, he had to concede that the tourism business was not classified by the federal government's production mobilization agency as a "defense industry," and therefore it qualified only for the lowest priorities in obtaining materials, supplies, and manpower.⁵⁶

Macnamee was also trying his best to harmonize travel with FDR's May declaration of a state of unlimited national emergency. Exaggerating, he claimed the president's action "sounded a call to the travel industry to serve the country in the nature of a second line of defense." He even adjusted the bureau's motto for the year to "Travel with a Patriotic Purpose."⁵⁷ USTB then refocused some of its travel PR. After all, not all travelers were leisure tourists. Rather, travel included the movement of men in military service, transfers of workers to new locations of defense-related factories, and families seeking to visit their loved ones wherever they may be.⁵⁸ Travel writers, who had a self-interest in maintaining the importance of travel, quickly followed this reorientation. The travel editor of the *Herald Tribune* noted that increased travel was partly due to "many thousands of relatives and friends of the men in [a military] camp devote their week ends or longer holidays to such visits."⁵⁹ The *Christian Science Monitor* noted the importance of travel by parents to visit their sons in military bases. It called such travel "inspection tours."⁶⁰

But reality was intruding on this Pollyanna thinking. In April, a riverboat cruise line in California announced the indefinite cancellation of all cruises. Its entire fleet had been requisitioned by the navy.⁶¹ By early May, a California hotel was shut down and taken over by the federal government for housing purposes.[62] Motorists on the West Coast were warned of significant army convoys between Fort Lewis (WA) and the Bay Area to occur during a one-week period in late May. Each formation might comprise up to 408 vehicles.[63] In October, airlines announced that due to sharp increases in passenger traffic, they were nearing maximum capacity. Given that air travel was "an important factor in national defense," they requested that travelers holding reservations they no longer needed cancel them as soon as possible so that the seats could be reallocated and not left empty at departure.[64] Defense jitters prompted the public's increasing concerns about how safe it was to travel at all. The bureau tried to respond with reassuring articles, such as how Washington's new National Airport included a control tower with the latest technologies to improve air travel safety and about detector cars constantly inspecting rail lines.[65]

By midyear, official statements from the bureau took on an increasingly desperate tone. A guest column from a USTB official argued that continuing to travel was "vital" for both economic and relaxation purposes.[66] Similarly, Macnamee wrote a column straining to make the case that increased arms production "does not mean necessarily" that individual workers would lose all their leisure time. It is "generally understood that frequent rest and relaxation periods from work are a necessity under present-day conditions, employers realizing that a fatigued and over-strained worker is an inefficient worker." Ipso facto, they can and should travel! In particular, these travelers "appreciate more keenly the calming and uplifting influences in National and State Parks and similar natural retreats."[67]

USTB realized that its messaging needed some kind of external endorsement from a higher authority and from someone with independent standing and credibility. Who was higher on the public administration food chain, short of the president? Paul McNutt. McNutt had been governor of Indiana and US high commissioner to the Philippines.[68] Returning home in 1939, FDR appointed him to head the new Federal Security Agency. It was an amalgamation of the many units scattered throughout the executive branch engaging in health, education, and welfare.[69] In late 1940, Roosevelt assigned McNutt a second role, to serve simultaneously as "the coordinator of health, medical, welfare, nutrition, recreation and other related activities affecting national defense" (Roosevelt, 1969, 1940: 529). (A reporter humorously noted that his clunky title was the longest and most unwieldly of FDR's defense appointments.[70]) This was a new position and was part of FDR's gradual effort to create a more flexible and responsive national emergency apparatus. Essentially, he was imposing over the legacy cabinet departments and tradition-bound agencies a fluid and adaptable exoskeleton. Within McNutt's new portfolio, recreation was relatively minor. It was the last specific subject listed in his cumbersome new title, suggesting that the preceding policy areas were more important. Also, the military retained exclusive control over recreation for all soldiers and sailors, a major constituency for recreation in 1941.

Nonetheless, McNutt had some vague oversight of recreation and travel by civilians. Therefore, Macnamee sought McNutt's endorsement of the bureau's promotion of tourism in the face of defense developments. At midyear, McNutt, ever the politician who hated to say no, obliged. However, it was with less-than-unconditional enthusiasm. He signed a statement for USTB that travel was "a potent stimulant to national health" and that it was "a worthwhile contribution to our home defense program." Leisure, he wrote, "must complement labor." For the public-at-large, travel was a "patriotic responsibility" because it contributed to "buttressing the nation in its hour of danger."[71] It wasn't a ringing endorsement, but beggars can't be choosy. It was better than nothing. And McNutt's standing gave the idea of travel-as-usual new validity. His statement was cited and repeated in later coverage in major metropolitan dailies.[72]

But real problems kept cropping up. Suddenly, the supply of gasoline for civilian use was getting very tight. This was a direct threat to automobile-based leisure travel, a mainstay of American tourism. In a release timed for the Sunday travel sections, USTB continued to encourage summer leisure travel, estimating that 17 million cars would be on the roads during that vacation period.[73] However, in late May, FDR designated Ickes as petroleum coordinator for national defense and delegated him virtually unlimited powers to oversee the industry (Roosevelt 1969, 1941: 196–98).[74] Ickes now had a built-in conflict of interest. On one hand, from his ongoing role as secretary of interior, he strongly supported USTB's mission. On the other hand, his new energy duties might compel him to discourage leisure and discretionary travel. Gas rationing and gasless Sundays were quickly looming as possibilities.[75] Ickes tried to proceed cautiously, hoping to straddle his gas-conservation and travel-promotion roles. One of his first actions was to ask all eastern governors (where the shortage was most noticeable due to limitations in pipeline and shipping capacities) to promote voluntary conservation of gas consumption by 20 to 33 percent. USTB, the travel industry, and AAA breathed a sigh of (temporary) relief.[76] The actual impact of a voluntary reduction in gasoline buying would hardly hurt their self-interest.

Even so, an auto club noticed the conflicting messages coming from Ickes's administrative empire and criticized it publicly. Here was Ickes urging gasoline conservation while his subordinate Macnamee was encouraging leisure travel.[77] A *Times* editorial flagged the inconsistency in a more lighthearted way. With USTB promoting continued vacationing and tourism, citizens would surely welcome the bureau's invitation to "defend our country by visiting fiestas and rodeos and viewing the wonders of the West." But with all the pressures for production increases and the economic mobilization, "one doesn't see how it can be managed. Our factories must thunder and hum, our offices clatter and buzz."[78] It was the journalistic equivalent of a shot across the bow. The fading of concerns about gas rationing were quickly replaced by another indirect threat to the travel business: a rubber shortage. In the summer, the Office of Production Management used its economic powers for a de facto takeover of the rubber industry. It promptly cut production of tires for privately owned cars.[79] There would be no more new tires once the current supply ran out. The industry and USTB again tried to soften perceptions that this war-related supply problem would cripple recreational travel. Retreads were just as good![80] In the meantime, car owners besieged suppliers for new tires before the current stock ran out.[81] Let the good times roll—even if on bald tires.

Another new defense-related problem impacting USTB was a bureaucratic one: office space. The bureau's choice location in downtown Manhattan was highly sought after by the many government agencies now doing defense-related business in New York. In August and September, the head of the New York office said it looked like they were about to lose the exhibit hall, back offices, and storage space. The federal Procurement Office quickly occupied those areas. The bureau's field office was only able to hold on to the front office and display windows. Even that was in doubt because the US Maritime Commission was talking about taking over its space *in toto*.[82] Indicating how serious the situation was, at the end of his August memo to Macnamee, he added a handwritten PS: "We'll hold on to what we've got for dear life."

In September, it was becoming increasingly clear that the administration's support for defense-related recreation was discernably different from what USTB was saying about the economic importance of the travel sector or the compatibility of leisure travel with the wartime mobilization. A presidential message to the annual conference of the American Recreation Association urged new recreation initiatives "in behalf of the young men in the armed forces of the nation, in behalf of defense industry workers and in behalf of the civilian population."[83] FDR made no mention of travel or tourism as important for the economy, nor did he encourage travel for its own sake. McNutt's address to the conference was similar. He raised no explicit objection to travel and tourism as aspects of recreation and morale, but he did not endorse it either.[84] Similarly, FDR's welcome message to AAA's annual conference two months later was subtly worded to discourage recreational travel without saying so. He acknowledged that car travel "is also a medium for recreation and pleasure. Obviously, in times like the present there must be readjustments . . . first attention must be given to road needs for defense."[85] Another subtle, but important, hint that travel was falling away from the defense agenda came when FDR reorganized McNutt's defense role in September. He changed McNutt's (second) title to Director of the Office of Defense Health and Welfare Services. While McNutt's duties continued to include recreation, it was no longer part of his title and that activity was barely mentioned in the executive order recreating the agency within the Office for Emergency Management (Roosevelt 1969, 1941: 369–72).

USTB soldiered on as best it could. It still had a role in promoting closer relations with Latin America by encouraging American tourists to visit South America and vice versa. It was still affiliated with the National Park Service and encouraged visits to parks and monuments as both patriotic and recreationally refreshing. The San Francisco office's newsletter for mid-October embodied these themes. It featured on its cover a photo of an army truck convoy bringing soldiers from Fort Ord to Yosemite National Park for a visit. One article praised V-Days, open houses for civilian visitors at Army camps as "another phase of travel's role in National Defense."[86] A longer article described USTB's involvement in "Linking the Americas."[87] In a speech in November, Macnamee unveiled another motto for the year, arguing that "travel anywhere in the United States is an adventure in democracy."[88]

Neither USTB nor its business constituency was willing to give up on promoting tourism as consistent with the national defense mobilization. They fought back as best they could. At an industry conference in Washington DC in October, Macnamee described enhanced efforts to promote travel as "a wholesome factor in the defense program."[89] He touched on promoting Latin American comity and "welding national unity."[90] But his main emphasis was the economic rationale that the industry wanted to hear: "The influence of travel in building the Nation's wealth lies in the jobs it provides for more than a million people a year, in its unique ability to keep money in circulation, pushing a stream of dollars to the farthest corners of the nation; and in utilizing the tremendous investments represented by hotels and all kinds of transportation mediums."[91] He had become the mouthpiece for a self-serving economic special-interest group, oblivious to the national agenda of the administration. A month before Pearl Harbor, USTB came up with a third slogan for the year in an attempt to mesh travel with defense: "See what you defend." The November issue of its *Official Bulletin* displayed the new motto at the top of every page and presented a new graphic insignia of it.[92]

∽

In summary, the unfolding events and changes in national priorities in 1941 (before Pearl Harbor) were a distinct threat to tourism and recreational travel. In the face of the defense mobilization, USTB's messages sounded like a bureaucracy seeking to justify its existence with increasingly lame excuses. It was like trying to fit a square peg in a round hole. The bureau was hanging on for dear life. It was not a defense agency (a designation eagerly sought by federal agencies) and was gradually being pushed off the administration's agenda. There was a subtle difference between what FDR and McNutt had said later in the year versus what Macnamee and the travel industry wanted to hear. In the context of the national defense mobilization and an unlimited state of emergency, the administration emphasized recreational activities only for those within that effort, such as recreation for soldiers and sailors and the physical health of defense factory workers. These two categories were explicitly part of the national emergency agenda. No one else. On the other hand, the travel industry was seeking its naked self-interest. It was economically dependent on leisure travel and spending and did not want to have its income and profits reduced due to the defense mobilization. It did not want to have to accept financial sacrifices in the name of the national interest. USTB and Macnamee had been coopted into being the industry's ventriloquist.[93]

7

Travel Promotion in Wartime?

1942–1943

The new reality of war dawned on USTB slowly and reluctantly.[1] During the first eleven months of 1941, it had largely been able to zig and zag around the real-world problems that were crowding the bureau's continuing promotion of travel. Sometimes slightly oblivious, sometimes deliberately misinterpreting facts and signals, for USTB, 1941 had been a very good year. There seemed to be little reason to expect 1942 to be otherwise. But the coming of the war to the United States suddenly overturned the political consensus that USTB had been carefully built upon active endorsement by the administration, support from private business for a federal role in promoting its economic prospects, and support from Congress for an executive-branch agency dedicated almost exclusively to public relations.

Public Relations Immediately after Pearl Harbor: What, Me Worry?

For USTB, in the first weeks after the Japanese attack on Pearl Harbor, one would not discern that Congress had declared war and that the country was now in combat. The first post–Pearl Harbor issue of *Travel West* was dated December 16, more than a week after the attack. It made no mention of the war. The only important organizational announcement in it was a note that, at the turn of the year, the periodical was about to shift to a monthly schedule, rather than its current semi-monthly issues. No reason was given, but this must have been planned before Pearl Harbor and was presented as a routine shift to take effect in the new calendar year, nothing else.[2]

Initially, from the perspective of USTB, the declaration of war did not seem to be a seismic event. After all, the bureau had functioned normally throughout 1940 and 1941, maintaining a business-as-usual approach vis-à-vis the news. It had weathered major international events in the European and Far Eastern wars in an unimpeded fashion. In fact, those developments had increased the potential for tourism within the United States. Similarly, USTB had worked its way through major domestic policy changes, including the establishment of the draft, the expansion of defense spending, the growing federal apparatus to control the economy, the declaration of an unlimited state of emergency, limits on travel capacity, and all other manner of the (nonwar) national mobilization for defense. During all these events, the bureau only had to slightly adapt its slogans and promotions to the newest conditions, but it never significantly reduced its basic message that travel was a good thing: whether for national morale, for patriotism, for deserved recreation, and for the economic vitality of the travel business. In that context, how different could it be now that the United States was a combatant nation?

That approach was initially how Macnamee sought to adapt to the congressional declaration of war. He put his best foot forward in a short column quickly distributed for publication in the first Sunday paper after Pearl Harbor.[3] He then wrote a longer version for the next Sunday. Despite the shock of Pearl Harbor, he tried to soft-pedal the impact of the war. He argued that, consistent with the new tight focus on winning the war, there was the "encouraging realization that travel is one industry which can continue, and perhaps even expand, without impairing our war efforts." He conceded that this claim "may seem startling and even paradoxical."[4] But, summoning all the arguments he could, he pointed to the precedents of the United States during WWI not discouraging domestic travel, to current UK policy permitting weekend trips, even to German wartime policies promoting internal travel during the war. He restated talking points from 1941 that travel was good for national unity and morale and that McNutt endorsed travel as an important recreational outlet for defense workers.

Then Macnamee made a relatively new protravel argument relating to macroeconomics of wartime. In 1940–41, USTB had often asserted that the travel industry was an important component of the national economy and therefore needed to be maintained for national economic health. Macnamee now argued that wartime travel was economically even more important because it was an outlet for consumer spending. He said, "Travel is encouraged by economic experts who are eager that the public, in addition to buying defense bonds, spend its money on recreation, on amusements, on services, on travel, on anything that does not adversely affect the arms program."[5] He was getting at the valid fears in Washington of inflation. With more people employed (or in the service), there was more money circulating in the national economy at the same time that major cuts were being made in production of consumer goods, such as cars, tires, appliances, and home construction. There could be too much money chasing too few goods, the formula for inflation. One of the reasons for savings bonds (besides funding the war's costs) was, as he correctly acknowledged, to absorb discretionary money away from personal spending for nonessential goods and services.

Macnamee was arguing that the same rationale was true for travel spending: that it was noninflationary because it was being spent on economic activities that were already being offered anyway, that travel was "good" consumption in a wartime economy. He was implicitly stating that the hospitality industry was one largely based on *fixed costs*: an existing hotel had rooms to fill (beyond those needed for defense). The same was true for railroad carrying capacity, restaurants, and so on. Spending on recreation was the *opposite* of spending on limited consumer goods. Consumption spending through travel was therefore not inflationary (and, furthermore, it increased federal tax revenues). This was a relatively sophisticated economic argument and likely was something of a stretch. Yes, the travel industry had some fixed costs, but it also had variable costs reflecting the level of business. In a country with looming manpower shortages, maintaining travel at prewar levels could entail pressures for increased staffing, thereby competing with defense and military manpower needs. That could also be inflationary.

USTB's PR apparatus quickly moved to distribute Macnamee's column as widely as possible throughout the travel industry, not just to newspapers for publication. For example, the editor of the *Official Bulletin* sent copies to stakeholders, such as to private tour companies.[6] Most likely, he mailed it to the *Bulletin*'s entire mailing list. Macnamee's column was depicted as a formal and authoritative statement of federal policy toward travel for the war.

The same day as Macnamee's column, Dunlap, the travel reporter for the *New York Herald Tribune*, wrote that "suggestions have been made" (by whom?) that Macnamee be appointed to the Board of Economic Warfare. Chaired by Vice President Henry Wallace, in the post–Pearl Harbor era, this super-cabinet agency was intended to be the fulcrum for all federal wartime economic policies, with power over the entire executive branch.[7] According to the article, the rationale for appointing Macnamee to the board would be "to co-ordinate the many branches of recreational travel for the duration of the war." It noted that, in the past, Macnamee "has frequently emphasized the need of such co-ordination to insure a proper flow of such travel in the interest of the physical and mental health of all the people."[8] The article was almost certainly based on a leak by Macnamee to the reporter. It embodied their mutual desire to maintain travel during the war and to make USTB even more powerful in that role. If appointed to the board, Macnamee would become the *de facto* czar of wartime recreation travel with power over other agencies and departments. It never happened—a reality check regarding the gap between USTB's ambitions and Washington's policy agenda and priorities for the war.

The January issue of the San Francisco office's newsletter, *Travel West*, tried to calm unfounded fears about travel in wartime, emphasizing that substantively not much had changed. Indicating that the message was having difficulty getting through, an article complained particularly that "the large cities on the West Coast have had to bear the brunt of this most deplorable practice in which timorous and unthinking people by the spreading and compounding of unfounded rumors [that] have not only created a threat to the vital flow of travel throughout the country, but also have struck a damaging blow to the National Morale."[9] As late as April, Bossemeyer, head of the San Francisco office, told a Montana statewide conference that tourism would and should continue during the war and that the state's travel industry may expect the war to have only modest impact on business.[10]

USTB continued pushing the story line of business-as-usual marketing not only in PR efforts through the media, but also in one-on-one outreach efforts. For example, an executive at the Santa Fe Railroad got a telegram in Ickes's name encouraging continued travel—especially to national parks—for morale and relaxation purposes. He read it aloud at a meeting of a California hotel organization.[11] A Chicago advertising executive asked USTB if national parks would be open in 1942. He received a letter from Macnamee replying that NPS did not plan to close any national parks (so far).[12] Macnamee's reply ricocheted around the travel press (perhaps amplified by a bureau press release) and was cited in the *New York Times* and *Christian Science Monitor*.[13] In early February, he was reported to have confirmed again that national parks would indeed open for the 1942 season as always.[14]

Meanwhile, during the first full month of the war, the travel reporters kept up the drumbeat that Macnamee was originating. They continued citing him as encouraging wartime travel.[15] This extended beyond the large metropolitan papers. For example, Macnamee's message was repackaged into tailored and parochial news by newspapers in Huron (SD), Portsmouth (NH), and Benton Harbor (MI).[16]

Secretary Ickes on Wartime Tourism, January–March 1942

On January 17, 1942, USTB issued a press release in Ickes's name endorsing "civilian travel for purposes of relaxation" when consistent with the priority of troop and materiel movement. It was something of an odd release because it contained no direct quotes from Ickes and several from Macnamee. The release was largely a summary of a "report" supposedly submitted to Ickes by USTB, much of which was the same as Macnamee's December 21 column.[17] Ickes's announcement was widely reported as the definitive statement of federal policy for the duration of the war. However, given that the United States had been in the war for only six weeks and that there was much major war news to report, the placement for this release was often buried in back pages, and the coverage was brief.[18] It was reported more prominently in trade magazines for various professions with a strong interest in the subject.[19]

Ickes apparently never reviewed the draft of the statement before it was released (perhaps because he was not quoted in it), nor did he appear to notice the coverage of it in any of the major metropolitan papers that he likely perused routinely.[20] The eventual editorial reaction was relatively extensive and largely arch, noting the contradiction between Ickes promoting continued travel versus his previous call—as federal petroleum coordinator—to citizens to reduce discretionary gas consumption and tire use.[21] However, these were published in relatively small papers that Ickes would probably not have seen immediately.[22] A few papers editorialized in favor of his statement because they were in states that had substantial tourism business.[23]

Using Clark's historiographic approach of maximum plausibility (2013, 48), the most likely explanation is that Macnamee had overstepped his authority and issued the January news release without explicitly clearing it with Ickes personally. And Ickes apparently did not notice the press coverage or negative editorial commentary. But then Macnamee arranged for the next issue of the *Official Bulletin* (backdated as the January–February 1942 issue) to trumpet the announcement front and center. The cover was a page-size photo of Ickes with the headline "Secretary Ickes Urges Civilian Travel for Relaxation to Aid Health, Morale."[24] That was followed by an entire page reprinting the January 17 press release. Another page presented a pastiche of press coverage of Ickes's January announcement, such as "Green Light Signal by Ickes to Boost Travel" from the *New York Journal-American* and "Travel as Usual Is Urged by Ickes" from the *Cleveland Plain Dealer*.[25] Somewhat oddly, the last page of the issue included an article about Ickes's testimony at a congressional hearing urging—in his capacity as the federal petroleum coordinator—the public to reduce gasoline consumption, including by eliminating unnecessary travel.[26]

The rest of the issue was an odd amalgam of traditional promotion of travel and defensive apologias for wartime tourism. Some of the usual articles trumpeted tourism, such as "Kentucky's Sesquicentennial," "National Sportsmen's Show," and "Rocky Mountain National Park." Other articles defensively urged continued travel during the war, including "Conventions and the War Effort," "'Budget Your Mileage' Is Wartime Slogan of American Automobile Ass'n," and "American Hotel Association Aids Activities in Victory Program." (In mid-March, an article in the Sunday travel section of the *Herald Tribune* copped material from one of the articles in that issue of the *Bulletin*.[27])

The new issue of the *Bulletin* began arriving in the mail in early March. (A congressman said his copy was delivered on Thursday, March 5.[28]) In short order, many newspapers reacted and with much more vociferous editorial denunciations compared to their January statements. For example, on March 8, a Georgia newspaper denounced the newsletter as a "whifflebat, which flies with its head where its tail ought to be."[29] That day, an Indiana newspaper denounced it because "paper is scarce and we are well aware that we may be criticized for wasting the piece upon which this item is printed but, on the other hand, these things are being done and we thought some of our subscribers might like to know."[30] The next day, an editorial in an Iowa paper sharply criticized him, both for the message conveyed in the issue and for the prominence of his portrait on the cover. Apparently from a nationally syndicated service of editorials (commonly used by small newspapers to save costs), the paper said the *Bulletin* was "a typical example of non-essential spending" in wartime. Besides "containing 20 pages of pictures and descriptions of national parks and travel subjects," even worse "gracing the cover of this publication is a large picture of Secretary of the Interior Harold Ickes." His endorsement of travel in wartime "is the usual attempt to associate the activities of a non-essential agency with the war effort."[31] The same day, the conservative *Manchester [NH] Union* condemned the contradictory messages from the federal government about travel as well as the fact that "publishing the bulletin isn't essential business, and its cost to the taxpayers, who foot the unnecessary as well as necessary undertakings of government, must be considerable."[32]

Over the rest of the week, the pounding continued. An editorial in another paper (also probably canned) talked about the confusion in Washington, particularly the phenomenon of high-ranking officials contradicting each other about national policy. This was the ultimate proof that Washington was "a veritable madhouse and no one knows heads or tails half the time . . . and all the while spending money by the millions."[33] A news story in a Texas paper noted the conflicting advice from "Doc" Ickes: Don't travel! Travel![34]

As soon as he began hearing about the issue and the press onslaught, Ickes hit the roof. According to his diary, he promptly summoned Macnamee to his office on Friday, March 13, and complained about the *Bulletin* putting him so prominently front and center as encouraging travel in wartime. "I didn't even know that the thing was being contemplated until newspaper editorials came in criticizing me," he claimed he said to Macnamee.[35] It is unclear *which* editorials he had seen by then. Some of the examples cited above might have arrived in his office by the 13th, though they were published in smaller and more distant papers. Presumably, some of them might not yet had been forwarded to him by the private clipping service he subscribed to. Most likely, Ickes was referring to an editorial in at least one of the major newspapers he routinely had access to and perused on a daily basis. It is also possible such an editorial was published in a lesser metropolitan newspaper and quickly forwarded to him by the clipping service of the Office of Government Reports (OGR). However, I did not find any editorials on the subject in the four major Washington papers (*Post, Star, Daily News,* and *Times-Herald*), two of the major New York papers (*Times* and *Herald Tribune*), nor in some other major metropolitan dailies, including the *Baltimore Star, Boston Globe, Atlanta Constitution, Chicago Tribune,* and *Los Angeles Times.* (After his meeting with Macnamee, more newspapers piled on in the same vein, sometimes in straight news coverage, other times in editorials.[36])

Ickes was a self-styled curmudgeon who never feared controversy for his actual opinions, such as the reaction to his AAA speech calling for trucks to stay off the highways on weekends (chap. 4). But he was particularly flummoxed because in this case he agreed with the newspaper criticisms of him. "I can't reply to criticism of this bulletin because it is justly to be criticized," he said. Very unhappy with Macnamee, on the spot, Ickes terminated any further publication of the *Bulletin*. Macnamee tried to argue back, saying that with all printing costs being done in-house, an issue only cost about fifty-five dollars. He also claimed (probably falsely) that for this issue "only a few had been sent out." Ickes was not mollified, saying that if just one "falls into the wrong hands, [it] could do as much damage as could a million."[37] As far as Ickes was concerned, USTB should be largely mothballed during the war, paralleling his decisions about reductions in operations of the National Park Service and other branches of the department not directly involved in production of raw materials or sources of energy.

At a mid-May news conference, President Roosevelt briefly addressed the issue of summer trips, thus conveying an authoritative, if lukewarm, policy of accepting some wartime travel for personal purposes. He said these annual summer vacations were okay as long as they did not have the effect of curtailing war-related travel, such as railroad and airplane seats. He said he would support new formal restrictions and rationing on such travel only if they came to be absolutely necessary for the national war effort. Otherwise, he would not interfere with or criticize such travel (Roosevelt 1972, 19: 329).

Organization and Leadership: Wartime Shrinkage during FY 1942

For the first few weeks after Pearl Harbor, USTB's operations did not reflect any particular change. Some routine organizational activities continued earlier patterns including publishing the semiannual *Calendar of Events* for the first six months of 1942 (USTB 1942). In terms of leadership, as of December 19, the *Congressional Directory*'s listing of the bureau's senior managers remained unchanged: chief Bruce Macnamee, New York Field Office supervisor Jay Wingate, and his San Francisco counterpart J. L. Bossemeyer.[38] Continuing the approach that nothing much had changed, the San Francisco office released the January issue of *Travel West*. While much of it was dedicated to reassuring reports about the continued availability of tourist sites and seating capacity with transportation companies, the issue informed readers that "plans are being developed for continuing the New York and San Francisco offices" and that the headquarters in DC "is being continued in full strength."[39] A later report from USTB said that both field offices would remain open but with "skeletonized" staffing.[40]

Seemingly out of sync with its own messaging, the wartime reality for USTB's operations, as opposed to its wishful thinking, hit home rather quickly after Pearl Harbor. During the last week in December, USTB abruptly announced that—contrary to its hopes—both field offices would close at the end of the year, only a few weeks away. As for the bureau's headquarters office in DC, it expected to shut down six months hence. Were it to continue at all after the end of FY 1942, this would have to underwritten in full by the travel industry.[41] However, this announcement about the bureau's fate was a bit premature.

Later in the spring, wartime realities were having a fuller effect. USTB was moved from its own offices in downtown DC to a small office suite in the Interior Department's building.[42] A conservative syndicated political columnist in April informed readers that "Ickes has closed his travel bureau and ordered his promoters to muzzle their typewriters."[43] He overstated it. A few weeks later, United Press accurately reported that USTB "is still functioning despite the war" in Washington, although without its field offices anymore.[44] The San Francisco and New York offices were quickly wound down by NPS in March and April. For example, New York's Wingate shipped his remaining stock of posters to the Washington office, transferred machines and office equipment to other federal agencies in New York, and NPS terminated the employment of the New York staff.[45] The rest of the office's holdings, collections, and records were stored in the space below the Statue of Liberty (an NPS site).[46] By late April, Wingate was alone in the New York office and reported to NPS that he was trying to complete a shutdown of the office by May 31. Also, the bureau's prewar assignment to promote South American travel "has necessarily been curtailed" (Lockwood and Smith 1941, 680). However, it was not terminated prematurely. During the first half of 1942, it fully expended the $25,000 Congress appropriated it to score twelve reels of travelogues about the United States in Spanish and Portuguese. They were then circulated throughout Latin America by the State Department (Ickes 1942, 164).

The updated edition of the *Congressional Directory* in late May reflected how much had changed since its December edition. It no longer listed Wingate and Bossemeyer. Macnamee was the only one still in it. Further, reflecting a wartime downgrading of the bureau, he was listed within the section on NPS's leadership rather than—as had been the case in the December 1941 edition—in a separate USTB section.[47] Ickes's official annual report after FY 1942 ended (on June 30) confirmed that the wartime work of USTB was sharply reduced, but that it was indeed still functioning. "A small force is continuing in Washington to serve as a liaison agency in supplying essential information to the travel industry, the war agencies, and to the public"

(Ickes 1942, 164). USTB's spending in FY 1942 conveyed this shrinkage. It will be recalled that Congress had appropriated it $75,000 for routine operations (chap. 6). This sum can be viewed as a spending ceiling. But due to limitations and realities that quickly came to impinge on the bureau after the war declaration, USTB barely spent half that—$42,000—before the end of the fiscal year in June 1942 (US House 1943a, A58).

Wartime Public Relations, December 1941–June 1942

During that first spring of the war, and Ickes's displeasure notwithstanding, Macnamee and USTB continued espousing the message of recreational travel as consistent with the war effort. He traveled, spoke, and engaged in media relations. In February, he was the dinner speaker at the annual meeting of the Atlantic City (NJ) Chamber of Commerce. He said, "this nation must maintain travel and vacation facilities for its people" so that workers could recreate and recharge their energies. He even urged continuation of the Miss America pageant and other celebrations "because they, too, afford enjoyable relaxation to maximum numbers."[48] A few days later, he spoke to an auto club in Washington, DC. Acknowledging that tire rationing would cause a cumulative reduction of travel, he nonetheless supported travel by club members.[49] At the end of February, he was in Manhattan addressing the first wartime meeting of the North American Travel Conference.[50] He and the organization endorsed travel and recreation consistent with war needs.[51]

A United Press reporter based in Washington called and asked if the early support for continued wartime travel and tourism was still the official federal policy in effect. A bureau spokesman (probably Macnamee) said, "We're standing pat on that." As an example of what USTB was actually doing, "it has been called upon recently to advise war agencies here as to where low-salaried employes [sic] might find suitable vacation resorts within 500 miles of the capital."[52] Later in the spring, the nearly empty New York office was still able to help arrange the annual Dude Ranch promotion effort in New York.[53] Commenting on apparent travel trends ever since the beginning of the war, Macnamee cheerfully announced, "We should promote travel to the limit in consonance with our war effort."[54]

During these first six months of the war, the economic self-interest of the travel industry and newspapers in continuing to promote travel under the protective shield of USTB was on naked display. Advertisements for ticket sales by the Santa Fe and Southern Pacific railroads quoted USTB's public statements of encouraging private citizens to continue traveling for personal reasons.[55] So did the tourist bureau funded by the southern California travel industry.[56] In July and August, the hotel association of Southern California bought ads in several newspapers saying that "in spite of rumors, Southern California hotel and resort life is normal." The hotel ads prominently featured a quote from USTB that during the war citizens "will need the wholesome tonic of recreational travel as one of the greatest maintaining forces of national morale."[57] Newspapers, eager to protect advertising revenue from their Sunday travel pages, also placed announcements that prominently cited USTB's message that recreational travel was patriotic. The *New York Times*'s headline was "Vacations Build Health and Morale," and the *Christian Science Monitor* dedicated a full page to an ad headlined "Vacations for Victory."[58] A promotional ad in the *Herald Tribune* for its upcoming special Sunday travel section on wartime travel similarly cited the bureau's encouragement of travel "as an aid to health and morale in war time."[59]

Congress Fights over USTB's Wartime Fate, Spring 1942

On January 5, 1942, President Roosevelt sent his first wartime annual budget proposal for FY 1943 to Congress. In it, he sought to cut USTB's annual funding from $75,000 in FY 1942 to $20,000 in FY 1943. He recommended full closure of both field offices, with the elimination of all twelve positions there. The president's budget document also recommended that the headquarters office be reduced to five people. After counting their salaries ($16,700), that would leave the bureau with only about $3,200 for all other expenses (US House 1942a, 676). It is probable that these recommendations reflected Ickes's political strategy. He had fought long and hard for a statute creating a travel bureau and formally placing it in Interior. He would not want to give that up totally and then have to refight the issue after the war. Instead, by severely cutting it, he could claim he was reflecting wartime priorities, while at the same time maintaining a skeletal Travel Bureau through the war years. Then, after the war, any political controversy would be about *how much* to restore to USTB's operating budget instead of debating *if* to recreate it at all and—most importantly to him—*where* it should be in the federal bureaucracy.

For the conservative coalition of southern and conservative Democrats and minority Republicans on Capitol Hill, the war presented a political opening to reframe their anti–New Deal ideology. Dubbed the economy bloc by the press, they could now attack New Deal agencies as simply expensive and nonessential during a war emergency. Here was a way to make their opposition to FDR's expansion of the executive branch look less ideological and possibly more appealing to the public. Who could be against cutting nonessential government activities during a war? It will be recalled that in 1940 a Republican member of Congress had denounced the alphabet agencies FDR had created and included USTB in his list (chap. 5). Now that war had been declared, another Republican engaged in the same line of attack. In February, Congressman Philip Bennett (R-MO) presented a similar list that included USTB. He asserted that at least two thirds of those agencies should be eliminated by Congress as nonessential to the war effort.[60] He picked a good time to do so. Spring was when the subcommittees of the House Appropriations Committee were just beginning to consider FDR's FY 1943 budget requests. Given that many of the agencies on his list were created by statute (again, including USTB), then the annual funding bills were the perfect platform for defunding them. He was drawing a line in the sand, particularly one that would appeal to the skinflints serving on the Appropriations Committee.

Two days later, the Interior appropriations subcommittee held a hearing on FY 1943 funding for the National Park Service. USTB was a prominent subject. Macnamee testified that the bureau's wartime goal was "helping to influence a rational and intelligent use of recreational facilities to maintain fitness for patriotic duty" (US House 1942b, 679). Argumentative subcommittee members attacked from both directions. Shouldn't the agency be zeroed out for the war? No, he said. With a small staff of five in the capital, the bureau would be able to coordinate between the policies of the war agencies and the travel business. Maybe the geography of the cuts should be turned upside down, another commented. Perhaps the coordination that Macnamee was describing should be done at the grassroots level? That would entail eliminating the DC staff where office space was tight and there were shortages of employees, but keeping the two field offices where office rental was less expensive and had a larger pool for potential staffers? No, said NPS Associate Director Demaray. They need to be where the war agencies are headquartered. NPS Director Drury tried to shift the framing of the issue. It was not really a bureau in the traditional executive branch meaning of the term for the largest major organization components of cabinet departments. Rather, it is merely "a travel information office" (681). Furthermore, it did not just promote visiting national parks, but it also provided information regarding all of Interior's travel destinations, such as Puerto Rico and popular engineering marvels such as dams.

Another committee member said he believed USTB duplicated functions with another federal agency. "I will not say exactly which—but a similar bureau somewhere in the Government" (684). No, said the testifiers. Another member suggested, given the staffing shortages in DC, "Could you not use these men who are now in the Travel Bureau in other branches of your department?" If the main activity of the remaining staff would be to reply to requests for travel information sent to Interior, "would you need four or five high-priced men to do that? You could get one good girl and get out all that answering" (684–85). Seemingly trying to rebut these attacks, two weeks later Congressman Clarence Lea (D-CA), sponsor of the original USTB legislation (chap. 4), submitted a written statement to the subcommittee on the importance of keeping the bureau going to serve authentic war needs (1071–73).

The hearing was a harbinger of the rest of the appropriation process. The ax fell a month later. The committee recommended roughly cutting the president's request slightly more than in half, to $9,820 (US House 1942c, 42). That figure was probably derived from a 50 percent cut in the president's recommendation minus further cutting an additional $180 for travel expenses. (Similar to printing costs, travel expenses were usually treated as a separate line item in budget documents, and the House Committee made a major effort to slash drastically all travel expenses of Interior's subunits.) If there was any merit-based rationale for the figure, then based on the president's estimate of five people at $20,000, it was probably meant to be just enough for two or three staffers—Macnamee, another lower-level PR specialist, a secretary—and administrative expenses. During the floor debate on the bill, Congressman Edward Rees (R-KS) proposed an amendment to defund USTB. Lea and Jed Johnson (D-OK), chair of the appropriations subcommittee, gave relatively full-throated defenses of keeping USTB (barely) alive, willing to face political attacks of wasting money on non-war expenditures. The amendment was defeated.[61]

The combative Ickes could have appealed the cut to the Senate Appropriations Committee, the routine role the committee played in the annual funding process. However, in the long list of issues he asked the committee to reconsider, he did not include USTB (US Senate 1942a). The most likely explanation is that he calculated the cut was minor compared to being wholly defunded. His major goal was to keep it alive during the war so that after the war it would still be a statutory and functioning bureau within Interior. In that context, it was a relatively minor detail whether it got $20,000 or $10,000 for FY 1943. However, Senator Carl Hayden (D-AZ), subcommittee chair, felt otherwise. He knew how important tourism was to his home state and was sensitive to the businesses there who cared about USTB. But he realized USTB was becoming something of a rallying point for the economy bloc. For example, two weeks before Hayden filed the committee report, Congressman Rees engaged in another attack on the bureau. In this case, he was not including USTB within a longer list of New Deal agencies. Instead, he focused on a handful of agencies (most engaged in publicity and information dissemination), including the bureau. "Why spend several thousand dollars for a United States Travel Bureau? . . . Why have a Travel Bureau when people are being asked to stay at home," he asked.[62] To avoid flagging this controversial agency, Hayden managed not even to mention USTB in the text of the committee report summarizing the recommendations for revisions to the House version. Instead, he buried it in a table of committee changes to the House bill, listing it as part of the committee's restoration of some House cuts in travel expenses for forty-one subunits in NPS. USTB was listed as regaining $180 in travel expenses and, by the way, $40,000 more for general operating expenses (US Senate 1942b, 16), i.e. $50,000 for FY 1943. It was two and one-half times what the president had recommended and five times what the House had approved.

The subject did not come up when the bill was debated on the floor of the Senate. On to the conference committee. In trying to develop a compromise version of the bill, the Senate conferees agreed to yield to the house position of $9,820, if the House would yield on restoring $180 in travel expenses, for a total of $10,000 (US House 1942d, 17). But the economy bloc was dominant in the House, and it rejected the conference committee report. In the floor debate, conservatives argued that the House conferees had yielded too much to the Senate. This was an argument that had institutional appeal, particularly given that the Constitution gave greater appropriation powers to the House than the Senate (a reminder that all funding bills had to start in the House). More specifically appealing to the conservative coalition, opponents argued the House conferees had given in to too much of the higher spending levels approved by the Senate. One of the issues they had raised was funding for Interior's "information items" that the Senate had sought to protect.[63] In a rare outcome, they won, and the conference report was rejected 169–143. Another conference committee was required. House conferees wanted to demonstrate to their members that they had gotten the message and the new compromise reflected more of the House version, especially in cutting Senate spending levels. While minor, this affected USTB, too. The Senate conferees agreed to drop their position of giving $180 to USTB for travel. They folded and accepted the House level of $9,820 (US House 1942e, 16). The bill was finally approved on the first day of the fiscal year, with the president signing it on July 2.[64]

Operations, July 1942–December 1942

Now USTB was down to three people, Macnamee, Mathilda C. Heuser, and Clara Henni.[65] This much-shrunk DC staff struggled on to promote tourism with the greatly reduced funding level. Not much. Most of the bureau's time was dedicated to internal federal coordination and planning activities with the war agencies, such as the army, the navy, the Office of Defense Transportation (a unit of the Office for Emergency Management in the Executive Office of the President), and Interior's Bureau of Mines and Solid Fuel Administration.[66] *Railway Age* reported that "work for the military services has occupied its staff in recent months."[67] For example, from its film collection, it provided the army with travelogues and other educational shorts to screen for soldiers. Using its photo collection of outdoor terrain, it helped army engineers design camouflage patterns. Military personnel posted overseas sometimes contacted USTB for travel information on the countries they were headed to. It also provided the State Department with visual materials about the United States for distribution in Latin America.[68]

The external PR activities it had routinely engaged in during previous years largely shriveled. They were few and far between. The bureau did not release any new publications or major press releases. In terms of media relations, it largely disappeared from the press and Sunday travel sections. One of the few occasions was when USTB was mentioned in a national wire service story about wartime vacation travel. Macnamee told a reporter that vacation travel was not diminishing due to the war. Actually, he said, the "war's tension and the fact that money is flowing freely have made people *more* vacation-minded than in former years."[69]

Another common activity in preceding years had been for Macnamee to travel to address out-of-town audiences. One of his very few trips that summer occurred when he traveled to western Maryland to assess its tourism potential, particularly because it was relatively close to the large population of civil servants and war workers in the capital. The idea came from the Baltimore & Ohio Railroad, always looking for more business to fill any excess capacity it had.[70] A local mayor took him around the county to see potential tourist points of interest. Macnamee said the area could be a year-round attraction, with lakes and fishing for summer sports, forests for hunting in the fall, and winter sports opportunities.[71] Winding up his visit, in a speech to a local civic group he said he was "sold" on the region's tourism potential.[72]

Denouement, December 1942–March 1943

Just three months after USTB's budget for FY 1943 had been signed into law, Interior was deep into its internal planning for the FY 1944 budget. By early winter, it would need to submit its budget plan to the Bureau of the Budget for review, revisions, and appeals. In October, Macnamee recommended shuttering USTB for the remainder of the war by not requesting any appropriations for FY 1944, that is, as of July 1, 1943.[73] His offer was quickly accepted, probably in part due to Ickes never forgetting a grudge or forgiving a slight. He had lost confidence in Macnamee at the beginning of 1942 because of the way Macnamee had put Ickes so front and center as recommending continued travel after Pearl Harbor. Now he had a chance to get his revenge. When President Roosevelt submitted his FY 1944 budget plan to Congress in early January, he indeed proposed zeroing out funding USTB (US House 1943a, A58).

The *Washington Post*'s columnist on federal civil service broke the story on December 9, 1942. In a sympathetic tone (perhaps because he liked Macnamee as an occasional source), he credited Macnamee with recommending eliminating his agency and his own job because its funds should be spent on the war. The columnist predicted that the shortage of office personnel in federal agencies meant that USTB employees would easily find new positions. He also used outdated information by stating that the bureau currently had one hundred employees.[74] The Associated Press quickly confirmed the facts with Macnamee and then moved the story on its national wire. It focused on the negative popular stereotype of public administration in the United States. The story's lead was "the antibureaucrat's dream—that a Government agency will suggest its own discontinuance—has come true."[75] It inaccurately stated that it had been Macnamee's own recommendation to Congress to cut his funding to $10,000 for FY 1943. The story got very wide play nationally because it was a kind of man-bites-dog tale.[76]

In general, the coverage depicted Macnamee as a kind of folk hero and patriot, which in reality was not the accurate inside story. Based on the fight to fund USTB in FY 1943, Ickes's loss of confidence in Macnamee, and NPS's lukewarm support for USTB given that its mission went beyond promoting national parks, Macnamee had simply seen the writing on the wall. Nonetheless, the press loved it. Headlines included "Anti-Bureaucrats Note: Agency Urges Own Discontinuance,"[77] "Page Ripley! Bureau Seeks its Own Finish,"[78] and "Wants U.S. Travel Bureau Ended: Director Would Save Funds for War Effort."[79] In a follow-up story the next day, AP confirmed with the Budget Bureau that USTB would go out of existence. It quoted an unnamed BOB official saying, "He will get his wish."[80]

Several newspapers editorialized their praise of Macnamee. A syndicated editorial, titled "Glory Be!" described him as "one of the most amazing persons" working for the federal government. "Did you ever hear the like, before?," it asked.[81] Another editorial was titled "Hail, MacNamee" and praised him as the Thane of Frankness. "His action, we are afraid, will ruin his reputation in Washington. Good bureaucrats there will look upon him with pity, pointing their fingers and then indicating sadly that a once bright man has gone completely wrong."[82] A few columnists praised him, too.[83] Other editorials grudgingly acknowledged his unusual action, but saw it as confirmation of their ongoing anti-big-government (and FDR) editorial stance. One pointed out that the idea of the federal government having a travel bureau at all "was unsound to start with."[84] Another said, "We'd like to know is why this Government agency . . . has been allowed to survive thus far into a period when Government agencies are frantically urging citizens not to travel."[85]

In October, when Macnamee recommended closing USTB, he hoped the three employees of the bureau could continue to work through the end of the fiscal year, if only to handle its shutdown in an orderly fashion.[86] But his recommendation to finish out the fiscal year on June 30, 1943, was not accepted by NPS (now headquartered, for the duration of the war, in Chicago). Instead, NPS set February 28, 1942, as USTB's termination date.[87]

Closing a federal agency required many details. First, the staff had to be formally terminated.[88] A freeze on further travel expenses meant Macnamee needed special permission from NPS to travel to New York in late December for two days to close out the bureau's contacts with the many travel business stakeholders based there. Looking to the postwar era, he wanted to be sure that USTB's industry partners did not feel insulted by abrupt closing of the bureau and then, due to this bad relationship management, refuse to resume a supportive role after the war. He thought that this considerate approach would "be helpful in bringing about a complete understanding of the elements which dictate the closing of the Bureau."[89]

His trip was approved. Well into 1944, NPS kept getting mail for USTB: requests for USTB publications,[90] to borrow slides and films,[91] to update mailing lists for press releases,[92] and from the US General Accounting Office (GAO) for a copy of the New York office's telephone contract for auditing purposes.[93]

Congress, Spring 1943

When Macnamee had proposed defunding USTB, his request was in the context of planning for FY 1944. The bureau's funding for the remainder of FY 1943 was already in place and was relatively modest, about $10,000. By spring of 1943, there were only four or five months left in the fiscal year, with less than $5,000 left to spend. Nonetheless, NPS made a firm decision to close the bureau no later than March 1, 1943. Why?

The most probable answer is that by March and April, Congress would be deep into the annual budget review for FY 1944. NPS apparently did not want to face a hostile appropriations subcommittee complaining about why USTB was *still* open. On the face of it, it would be a relatively reasonable position for NPS to state that it was only three people, that it had less than $5,000 more to spend, that closing a government agency needed to be done in an orderly and planned way, and that—anyway—the president's budget request already had been to zero out the agency for the next fiscal year. But NPS did not want to be in that defensive position at a public hearing. After all, in politics sometimes symbolism is more important than rationality and fairness. Better to put the issue behind it.

Indeed, that is how the issue played out. On April 6, 1943, the Interior Subcommittee of the House Appropriations Committee held a hearing on the FY 1944 budget for NPS. Right off the bat, NPS director Drury began his presentation by emphasizing how NPS had "curtailed its activities and adapted itself to war conditions" (US House 1943b, 791). In his prepared statement (that would have been distributed to subcommittee members at the beginning of the hearing), he said that "activities like the United States Travel Bureau . . . are being discontinued" (796). To be sure that the message was received, a few minutes later, Drury verbalized it: "The United States Travel Bureau has been abolished" (794).

But that wasn't enough for Jed Johnson (D-OK), the subcommittee chair. He was still stewing about a bureaucrat having seemingly slighted him. Johnson had felt burned and insulted by the press coverage lionizing Macnamee for recommending closing his own bureau.

> I notice he broke out in the press a few weeks ago. His name was on the front page of almost every paper in the country. They said, "Here is one Government official who voluntarily requests that his agency be abolished.["] He was held up as a great hero. The fact is, if I remember correctly, that after this committee had refused his agency an appropriation last year he slipped over to the Senate and urged that it be restored; there is just one reason why it was abolished, or at least the main reason was because this committee felt that he should not have the money to keep it going. I do not personally appreciate his posing as a great hero. (794–95)

The ranking minority member, Albert Carter (R-CA), promptly endorsed Johnson's comment: "I think the chairman of this committee should take credit for that. It was the chairman who abolished it" (795).

This brief snippet is false on several facts. First, in the spring of 1941, the House committee had not recommended zeroing out USTB for FY 1943. It recommended $9,820 rather than the $20,000 requested by the president. Second, Secretary Ickes did not include in his appeals to the Senate Appropriations Committee a request to increase USTB funding. He never brought it up and Macnamee did not testify either. Jones mistakenly remembered that Macnamee had testified before the Senate committee the year before that (1940), regarding FY 1942 (chap. 6). Third, in conference committee in June 1942 (which Johnson *chaired* [US House 1942d, 1; 1942e, 1]), the House conferees did not try to zero out funding for USTB. Doing so would have violated the conventional approach by conference committees, namely, finding a compromise between the positions adopted by the two houses, and to avoid injecting an entirely new one. Instead, the Senate ultimately yielded to the House position (a difference of $180 for travel expenses). Johnson's comment reflected how the world looked to people in his position. He seemed most agitated that Macnamee got public and political credit for closing USTB. As a politician, he wanted it—even if getting that credit was factually false. In politics, appearance trumps truth. Johnson was probably reflecting the sense of self-importance that congressional appropriators had, how much elected officials craved good press, and the traditional conservative view of public administration, that agencies should be seen but not heard (Lee 2011). Certainly, bureaucrats should not be the good guys at the (implicit) expense of politicians.

On March 1, 1943, Macnamee confirmed to a former NPS colleague, "I have closed the U.S. Travel Bureau."[94] That day, he began working for Pan American Airlines in its PR office in Washington, DC.[95] A few months earlier, when he had recommended closing USTB for the war, Macnamee looked back and felt that it had accomplished a lot in its brief existence. He said, "The Bureau had attained the main objectives for which it was created. It had firmly established its position as the Federal liaison for the travel industry of the United States. It had won the support of all the major segments of that industry. It had become the clearing house in all travel matters involving the other American Republics and the Dominion of Canada."[96] The travel reporter of the *Christian Science Monitor* agreed. He felt the agency was "comparatively new and already helpful." He regretted that the federal government had closed USTB, unlike Canada and Mexico, who kept their tourist offices open.[97]

8

Postwar Revival

Interior's US Travel Division, 1946–1949

After the National Park Service closed USTB in early 1943, it consistently stated that the bureau had been defunded due to the war, thus emphasizing the desire for resuming its operations after the end of the war. For example, replying to a letter from the Cuban tourism office in 1943, NPS acting director Tolson wrote, "It is hoped, however, that the Bureau can be revived after the war in some way as there is a need for some such agency in the Federal Government." However, remembering the legislative wars over which cabinet department had jurisdiction for travel, the letter carefully noted that "it is impracticable at this time to say just what form a revived travel bureau should take, or in what Federal agency it should be located."[1] On another occasion, at a congressional hearing in March 1945 (so before VE-Day), NPS's Tolson said that any postwar plans for it to reopen USTB would not occur "unless the committee gives appropriations for it" (US House 1945, 704).

Macnamee had been thinking along the same lines. A few weeks after closing USTB, he wrote an NPS official that he hoped "the Bureau, when peace is restored, will reopen and will have a brilliant future."[2] Similarly, only weeks after VJ-Day, the travel reporter for the *New York Times* began speculating "whether or not a United States Travel Bureau should be re-established, and if so on what basis." She even proposed its first project, a promotional campaign of "Victory Vacation Year" for 1946.[3]

The Democratic Seventy-Ninth Congress Declines to Revive USTB, 1946

However, the timing was awkward politically. FDR died in April 1945 and was succeeded by Vice President Truman. In the name of continuity, Ickes stayed on and, among his many other longstanding policy interests, was hoping to revive USTB. But in late 1945, when BOB was developing the president's budget recommendation for FY 1947 (to be submitted to Congress in January 1946), it recommended that refunding of the bureau should be put off at least one more year (US Senate, 1946a, 1315). At the same time, Commerce was trying to assume the postwar lead role and squeeze Interior out. In April, Secretary of Commerce Henry A. Wallace wrote a guest column for the Sunday travel section of the *Boston Globe*. He said the department considered the business of travel so important that "we are now in the process of organizing a commercial and tourist travel division." He emphasized that there was a need not only to encourage international travel to and from the US, but also to encourage Americans to "travel within our own borders."[4]

Almost predictably, the argumentative Ickes had a falling out with Truman and resigned in February 1946. The president nominated Julius A. Krug as Ickes's successor. In March, Krug was confirmed by the Senate and sworn in. By now, it was quite late in the congressional consideration of the FY 1947 Interior budget. The House Appropriations Committee had reported out the department's budget bill in early May (with major postwar cuts), and a (somewhat) softened bill was approved by the House in mid-May. Traditionally, the secretary would now appeal some of the House cuts to the Senate Appropriations Committee. Those appeals usually related to changes the House had made to the president's original January budget plan. However, Krug was casting about for some modest policy initiatives that would be *his*, rather than merely a continuation of Ickes's budget plan. To succeed, this meant he would have to ask the Senate Committee to step outside its traditional role and agree to consider new subjects, not just appeals from House actions.

Krug submitted to the Senate Appropriations Committee a supplemental budget request for NPS containing three items, one of which was the revival of USTB. According to associate NPS director Demaray, "when Secretary Krug came in, he immediately took the matter up with the Bureau of the Budget, as he felt now was the time to prepare for the future" (US Senate 1946a, 1316). As a result of his personal intervention, President Truman and BOB approved it and, in a formal message to Congress, requested (among many other items), $38,500 to start up USTB in FY 1947. The funding was intended to cover hiring a small core staff who would "assist the advisory committee in developing plans and programs" for FY 1947 and beyond (US Senate 1946b, 4).

Indicating that the pre-war alliance with business remained strong, when Demaray testified before the Senate committee, he said, "Yesterday representatives of the United States Chamber of Commerce came to my office and were very much concerned to find that funds had not been provided for the Travel Bureau . . . and they felt that the Park Service was the agency that could best serve their interests" (US Senate 1946a, 1316). The committee approved the request (US Senate 1946c, 37), as did the Senate. But the House would not yield; not even willing to compromise by splitting the difference to half that amount (US House 1946, 31). Not a cent for USTB for FY 1947.

The Republican Eightieth Congress Revives USTB, 1947

The November 1946 elections were a major win for the Republicans. Some of the Republican campaign themes were the need to rein in big government as created by FDR and maintained by Truman, cut government spending, and the need for the federal government to be more probusiness. They won the majority in both the House and Senate. (This was the Congress that Truman later dubbed the "Do-Nothing 80th Congress" in his 1948 campaign.) Undeterred by this partisan setback, Secretary Krug maintained his enthusiasm for resurrecting USTB. Krug's departmental budget submission to the Bureau of the Budget in late 1946 was for $100,000 for USTB in FY 1948 (US House 1947b, 553). Given that BOB had already approved, in principle, the resurrection of USTB in the supplemental presidential budget message to the Senate for FY 1947 (see preceding subchapter), the president's budgeteers retained it but cut the secretary's request by 25 percent. The president's budget proposal submitted to Congress in January 1947 included $75,000 for USTB to fund eleven staffers (US House 1947a, 431, 551).

At the hearing by the Interior Subcommittee of the House Appropriations Committee, majority party members wanted more explicit information about what, exactly, USTB would do. NPS officials were at pains to emphasize some details that would be important to Republicans. USTB would promote domestic travel to all locations in the United States, not just national parks.

This would include facilities maintained by other federal agencies, such as USDA's Forest Service, as well as state parks and other nonfederal destinations. The bureau would not compete with for-profit travel agencies, rather would be "a fact-finding and correlating agency" (US House 1947b, 553).

One majority member focused on the information that USTB would disseminate. Was this a "propaganda service?" The subject of federal agency PR was a key across-the-board concern for Republicans. In fact, the Eightieth Congress created a special committee to investigate all government PR programs (Lee 2011, 169–73). NPS representatives emphasized that USTB materials would not be propaganda and would be consistent with the kind of information it disseminated about parks. Furthermore, bureau public relations programs would reflect "the total industry," including distributing to the public information based on "guidance that the industry would want." And it would not take sides in any "regional competition" (554). NPS also emphasized that during the restart-up year, USTB would create an advisory committee, to include industry and state officials to "recommend a program" that USTB would then implement (615).

The subcommittee recommended full funding of the president's budget request for USTB (US House 1947c, 47), even though the House made major cuts in departmental funding.[5] This indicated the unusual place that the bureau occupied on the political spectrum. Business support for USTB turned out to carry greater weight than Republican doctrine to cut all government spending and reduce all informational programs. That the bureau was a (somewhat) new bureaucracy, new spending, and new hires that expanded big government were not fatal problems. Then the Republican majority in the Senate accepted the House recommendation for USTB. Major ideological fights over larger spending issues of the department delayed congressional passage of the bill into the beginning of the fiscal year. However, through all those controversies, USTB was never challenged. On July 25, 1947, Truman signed the FY 1948 Interior appropriations bill, including $75,000 in new spending and new personnel for USTB.[6]

The US Travel Division's Operations and PR, 1947–1949

Secretary Krug moved quickly to reestablish the office. In August, he began the process of reactivating the statutory advisory committee by inviting six federal agencies with relevant activities to appoint representatives to sit on it.[7] The committee would also have an equal number of industry members.[8]

Behind the scenes, Krug offered the job of heading the nascent unit to J. L. Bossemeyer, who had headed USTB's San Francisco field office before the war.[9] (When the office had closed in 1942, Bossemeyer transferred to NPS's regional headquarters in San Francisco and worked there during the war as one of its three park planners.[10]) Reminiscent of the earlier practice of USTB (chaps. 2–4), Krug also appointed three men as "collaborators," akin to expert consultants, to assist in the office's resumption.[11] Later, three more were added to serve in that role.[12]

On Saturday, October 18, 1947, the National Park Service issued a press release with Secretary Krug announcing officially the reopening of the office, retitled as the United States Travel Division (USTD).[13] Travel reporters joyously wrote about the news in their columns on the travel pages of the Sunday papers.[14] Krug had timed his announcement so that Bossemeyer could quickly be scheduled as a speaker at the annual conference of the National Association of Travel Officials, set for October 23–25 in Chicago.[15] Bossemeyer's major news when he addressed the group was that the new USTD would focus exclusively on promoting domestic tourism. Breaking somewhat with the agenda of the prewar USTB, USTD would not try to encourage war-ravaged Europeans to visit the United States, nor would it seek to encourage US tourists to visit Latin America.[16] Nonetheless, in 1948 it gradually resumed USTB's pre-war mission of promoting travel to the United States by foreign citizens, including printed materials to be distributed abroad.[17]

It is unclear why Krug changed the name of the agency. It is possible that the term "bureau" suggested a hierarchal status, based on the conventions of the executive branch at the time. Generally, bureaus were the building blocks of cabinet departments, such Interior's Bureau of Reclamation and Bureau of Mines. The confusion may have been compounded because it was within the National Park *Service*, itself the administrative equivalent to a bureau. It is also possible that the name change was a prelude to moving the unit out of NPS, particularly because the industry stakeholders chaffed at the implication that the core mission was to promote the national parks (see next subchapter). (It will be recalled that, in 1940, Secretary Ickes had briefly moved the unit from NPS to the Office of the Secretary when he thought that might help get the bill passed by Congress. It didn't [chap. 5].)

The revived travel agency was a much more modest operation compared to the prewar USTB. Unlike its predecessor, USTD was limited to an office in the Interior building in Washington and had nine employees (US House 1949a, 333). It did not have any field offices. Also, the DC office was not designed to have a walk-in service counter for visitors seeking information. Instead, all public inquiries were handled by mail. In-person visits were limited to reporters, researchers, other federal officials, and travel professionals (USTD 1948, 2). The immediate plans for the division were to work with the advisory committee to develop a plan of action and to reestablish prewar working relations. Gradually elaborating on the immediate agenda of USTD, one goal was to encourage year-round tourism instead of the traditional summer travel.[18] For example, national parks were at full capacity with families in the summer but underused and less crowded during the school year. USTD encouraged tourists who did not have school-age children to visit the parks during the other three seasons.[19]

Another goal was Bossemeyer's sense that there was a need for more research and reliable statistics on travel, particularly its impact on the economy. In 1949, USTD published a brochure summarizing available information on the industry but acknowledging that these statistics were "inadequate" because they were often estimates and not grounded in rigorous research methodology (USTD 1949a, v). In part, the publication was intended to help make the case for the federal government to collect authoritative travel statistics, just as it did for so many other aspects of the national economy. USTD was even hoping that some questions about travel could be included in the upcoming 1940 census. However, that appeared unlikely. That the Census Bureau was in the Commerce Department, the loser so far in the turf fight over ownership of USTB/USTD, probably didn't help.

The most significant difference in missions between the USTB and USTD was an entirely new policy goal of conservation. USTD sought to cooperate with and encourage programs "designed to conserve the scenic and recreational resources of the United States" (USTD, 1948, 5). In part, this new policy goal could be seen as helping integrate USTD into some non-NPS components of the Interior Department, such as the Fish and Wildlife Service. It also created a bridge for relationships with other federal agencies, such as USDA's Forest Service and its soil conservation program, as well as with the forty-eight state park systems. This new policy goal also included two remarkable subjects. First, it sought to promote "protection and proper marking of historic and scientific places *not* in park areas" (emphasis added). This expanded the agency's scope of interest beyond the confines of formal parks. Second, intriguingly, USTD now sought to promote conservation goals of "prevention of stream pollution" and "roadside protection and beautification." This was prescient. Interest in pollution and beautification did not become significant public policy issues until the 1960s and beyond.

In many respects, USTD's public relations to promote travel picked up where USTB had left off. In mid-1949, it resumed publishing a national calendar of events of interest to travelers (USTD 1949b). It restarted a monthly magazine "written by travel experts for travel experts" with a circulation of five thousand copies (US Senate 1949, 345). Initially called *Travel USA Bulletin*, the first issue was released in October 1948.[20] More sophisticated in presentation than the *Official Bulletin*, it was typeset, included pictures, and had some color. In January 1949, USTD shorted the title to *Travel USA* and continued the monthly publication schedule.[21] (The last issue, the twelfth, came out in September 1949.) USTD then began publishing a new series of publications titled *Sources of Travel Information*. The four issues in the series were *Travel Information Offices Maintained by States and Territories* (no. 1), *Where to Get Information about State Parks and Recreational Areas* (no. 2), *Touring Services for Motorists Offered by Gasoline and Oil Companies* (no. 3), and *National Forests and Parks: Federal Recreation Areas* (no. 4).[22] It also published and distributed free upon request a large color map of *Recreation Areas of the United States under Federal or State Administration*.[23]

Like his predecessor, Bossemeyer traveled to give talks to conferences and at other public events. This not only gave him a chance to plug the division's activities to its stakeholders, but also to garner newspaper coverage to promote awareness of USTD to the public-at-large. For example, he spoke at a conference on tourism at Washington State College in Pullman (WA),[24] the annual conference of the American Society of Travel Agents in Savannah (GA),[25] the annual conference of the National Association of Travel Officials in Miami,[26] a conference of the Northern Great Lakes Area Council in Blaney Park (MI),[27] and the Mississippi Valley Association's meeting in Illinois.[28] USTD's assistant director, Ralston B. Lattimore, was one of the VIPs at the formal dedication of the national military park in Chattanooga (TN).[29] Other PR activities included sponsoring an exhibition of travel advertising art,[30] encouraging local studies of the economic value of tourism,[31] and arranging for a full-page article and mail-in coupon in the national Sunday newspaper supplement magazine *This Week*.[32]

The Republican Eightieth Congress Continues Funding USTD but Declines to Elevate It out of NPS, 1948

President Truman continued supporting Secretary Krug's effort to upgrade USTD. In January 1948, he recommended to Congress to increase its budget from $75,000 in FY 1948 to $90,000 for FY 1949 (US House 1948a, 555). At the House hearing, Congressman Ben Jensen (R-IA), chair of the subcommittee, seemed less impressed by the support for USTD from the industry than so many of his other probusiness colleagues. "It is not clear in my mind yet just what they do," he said. Even after NPS named the endorsements for the division from national travel organizations, he still said, "I am trying to find out what the purpose is" (US House 1949b, 688). Reflecting the competing political siloes of congressional Republicans who were probusiness versus those who wanted to cut spending, the House rejected Krug's increase request and instead cut funding to $60,000 for FY 1949 (US House 1949c, 63). The Senate was more supportive, with the chair of the full committee indicating support for the president's request (US Senate 1948a, 504–05). Eventually, the committee recommended keeping funding at the current level of $75,000 (US Senate 1948b, 32, 53). However, the House insisted, and the Senate receded (US House, 1948d, 30). In late June, Truman signed the Interior funding bill into law giving USTD $60,000 for FY 1949.[33]

Secretary Krug treated the just-approved appropriation as a news peg for a second public kickoff of USTD. In June, he announced the "reactivated" unit based on recommendations from a twelve-person advisory committee. His actions provided "the groundwork for a national travel office for the development of travel in the United States and to its territories and island possessions." However, the current level of funding would not cover the costs of a full panoply of agency activities; rather, it would permit "only the most essential portions of the program."[34]

A stronger indication of how much the agency had coopted the travel industry (or vice versa) occurred on a different congressional track in 1948. By now, the industry considered the office to be *its* agency and representative in the federal government. The importance of USTD was perceived as a synecdoche for the importance of the business itself. As an entity buried within NPS, USTD looked unimportant. Furthermore, its location within NPS continued to irk those who sought for it to be independent of any role vis-à-vis parks. They wanted their *own* agency, a freestanding federal office. For example, the AAA wanted USTD "raised to the status of a Bureau of the Interior Department."[35] Similarly, the National Association of Travel Officials wanted USTD to be "a full-fledged separate bureau" of Interior.[36] The travel reporter at the *Chicago Tribune* also called for "bureau status for the recently created United States travel division."[37]

Reflecting this, in April 1948, Congressman Charles Wolverton (R-NJ), chair of the Interstate and Foreign Commerce Committee introduced H.R. 6136. (The committee's ranking minority member was Clarence Lea [D-CA], who had been a major proponent of the original legislation.) The bill proposed to move the unit out of NPS and recreate it as a freestanding bureau in the Interior Department, bureaucratically equal to the department's other major components. Its name would be United States Travel Bureau. The bill also eliminated the $100,000 cap on authorized annual funding that had been in the original 1940 statute. Instead, Congress could decide on the level of funding as it "may deem necessary to carry out the purposes" of the bureau. Given that the bill had been introduced by the committee's chair, its prospects were good. Wolverton quickly convened a hearing on it in May. It was something of a love feast for USTD/USTB, with praise coming from the National Association for Motor Bus Operators, Air Transport Association of America, National Federation of American Shipping, and National Association of Travel Officials. All endorsed the legislation and testified to the importance of the unit being an autonomous and freestanding bureau in Interior.

There were only two discordant notes at the hearing. In a letter, Commerce Department secretary Charles Sawyer opposed the bill because the department currently engaged in support for international travel. Given the "present policy which directs the Department of Commerce to take the lead with regard to international travel, any sizable extension of another Department into this sphere would seem inappropriate." Furthermore, "we are concerned with the possibility of duplication of activities undertaken to encourage travel in the event of expansion of such activities by other departments" (US House 1948e, 4). Committee member William J. Miller (R-CT) opposed the bill, suggesting it "offered a case history of how government costs expand."[38] As a conservative and opponent of big government, he wanted to stop this agency before the typical bureaucratic thickening process set it. Yet his stance was ideological, while the other majority party members viewed the bill as probusiness.

Two weeks later, the committee recommended the legislation for adoption. Highlighting the bill's bipartisan support, Wolverton permitted minority member Lea to be listed as the submitter of the committee report (US House 1948f). The bill passed the House in mid-June without debate.[39] Arriving in the Senate, it was referred to the Interstate and Foreign Commerce Committee and never heard from again.[40] (Another effort by House

Republicans to enact such legislation in the 1949 Democratic Congress failed even to get a committee hearing.[41])

The Democratic Eighty-First Congress Kills the Travel Bureau, 1949

The ground shifted during the winter of 1948–49. In the stunning and unexpected upset in the November 1948 elections, President Truman won, and Democrats regained the majority in both houses of Congress. Then, in early January, a task force of the Hoover Commission, an ostensibly nonpartisan and disinterested expert review of public administration in the federal government, recommended "discontinuance" of USTD because private businesses and nonprofit associations "issue a vast volume of travel information which fully informs the public respecting travel opportunities in the United States" (US Commission on Organization 1949, 62). It was a good government and efficiency argument that added weight to the political opposition. This recommendation gave credibility to politicians who opposed the existence of a travel bureau whether because they saw it as corporate welfare or as an example of FDR's big government.

With Democrats in control of the Eighty-First Congress, the argument that the Travel Division should continue to be funded because that was what business wanted carried less weight than it had had vis-à-vis the Republican majority of the Eightieth Congress. Now the obligation fell on USTD supporters to overcome the skepticism of the penny-pinching skinflints (whether Democrats or Republicans) of the House Appropriations Committee to keep funding it. They were not persuasive. In March, the committee recommended zeroing out USTD's budget for FY 1950 and explained its rationale with this very strong statement: "The committee considers it an utter waste of funds for the Government to promote and encourage travel. Commercial transportation interests spend tremendous amounts for such purposes" (US House 1949b, 19). In the floor debate, a Republican from Atlantic City (NJ) tried to restore the funding. The subcommittee chair said Americans preternaturally travel and don't need the feds to encourage them to do so. The amendment to fund USTD failed, 22–76.[42]

Secretary Krug made a last-ditch and personal effort to get the funding restored by appealing to the Senate Appropriations Committee. He said, "Many thousands of business firms throughout the Nation, dependent on travel for part or all of their business, will not agree with this conclusion" of the House (US Senate 1949, 32). NPS's director commented on the House report saying, "I think, however, there has been a misunderstanding as to the true nature and function of the Travel Bureau" (293). While the committee eventually declined to recommend restoring USTD's funding, a Senate floor amendment to do so (at $63,600 for FY 1950) was accepted with little debate and no opposition.[43] Yet again, the House conferees (who were the senior members of the Appropriations Committee) were unwilling to compromise. Eventually, the Senate conferees yielded and the bill provided no funding for USTD (US House 1949c, 21). When the bill was finally sent to the president in October (due to disagreements over larger matters[44]), Truman's signature ended the federal travel program.[45] The office closed so abruptly that only five years later, a Commerce Department official testified before Congress that he didn't know of any extant document recounting the experience and accomplishments (or lack of) by Interior's USTD (US House 1954, 14).[46]

Bossemeyer formally announced the closure of USTD in November 1949.[47] Reflecting the subject they covered, travel reporters lamented what Congress had done. One criticized "congressional indifference" for an office that "only recently has begun to get into full swing a program of encouraging travel by every one [sic]."[48] The Sunday travel page of the *Chicago Tribune* reported that USTD had done "an excellent job for the travel industry."[49] The *Post*'s travel editor wrote that it was a "sad little note" to have to report on USTD's closure.[50]

The funding process had zeroed out any spending for USTD by 1949. Case closed and end of story? Was Interior's travel promotion mandate dead and gone? Not quite. The bifurcated nature of congressional decision making left an important loose end. The 1940 law creating USTB and authorizing funding for it was still on the books. The appropriations process could not touch authorization statutes, only funding matters. To erase Interior's PR mission to promote domestic travel, the 1940 law would have to be repealed. That could only happen through a separate (nonappropriation) bill and through the authorizing standing committees, the commerce committees of the two houses. They were wholly independent of the appropriations committees.

9

Last Try

Interior's Office of Travel, 1968–1973

It did not take long after the closure of USTD in 1949 for the business, bureaucratic, and media constituencies to resume calling for the federal government to have an office to promote travel, that is, to benefit their overlapping self-interests. The key difference between the flameout of USTB and these new efforts was that almost every suggestion involved locating it in the Commerce Department. The rationale and political push, beginning with Ickes in the 1930s, for it to be in the Interior Department had largely evaporated and disappeared. But the 1940 law had not.

Travel Promotion Policy and Politics: Presidents Eisenhower and Kennedy

In the immediate postwar years and particularly after the defunding of USTD, most of the federal government's interest in travel was related to the reconstruction of Western Europe. Parallel to the Marshall Plan, increasing American tourism to Europe was viewed another way to transfer dollars to European countries to help finance reconstruction. There was even a program office in the federal government's Economic Cooperation Administration (ECA) to assist European governments in restoring their tourism attractions and facilities so as to draw more American visitors.[1] Due to this Eurocentric motivation, there was no reciprocal federal attention to encouraging foreigners to visit the United States nor of promoting domestic tourism. Gradually, the fiscal situation turned upside down. The United States began having a balance-of-payments deficit, partly because so many American tourists were visiting Europe. To Washington politicians, the problem was now that not enough foreigners (particularly Europeans) were visiting the United States. "Do something!" was the sudden demand on the federal government regarding international travel, of enticing more foreigners to visit the United States.

During the first year of the Eisenhower administration, the Commerce Department created a small Office of International Travel to encourage foreign travel *by* Americans.[2] Throughout Eisenhower's two terms, it was a minor bureaucracy, with annual budgets in the range of $65,000–78,000 and about four to five staffers (US House 1954, 6; US Senate 1960a, 28; US House 1960, 37).[3] However, the mandate of the agency was increasingly depicted in ambiguous terms, such as "to stimulate and facilitate world travel," vaguely implying that tourism promotion included increasing visits by foreigners to the United States as well.[4] Inevitably, the constituency that would benefit from increased travel to and in the United States became increasingly vocal and insistent toward the end of the Eisenhower presidency. The travel editor of the *New York Herald Tribune* wrote of the "Need Seen for a U.S. Tourist Bureau," and his counterpart at the *Washington Post* applauded that a "Program to Increase Travel Grows."[5] By now, the advocates were able to turn the postwar policy upside down. There was an urgent economic need to plug the dollar gap, to reverse a net outflow of dollars that was contributing to the trade deficit.[6]

The industry wanted its "own" bureaucracy and congressional allies were glad to draft something to their satisfaction. Reflecting that, a White House economic advisor said that the current bureaucratic status of the travel office "suffered from lack of emphasis and stature," particularly because of its low standing in the bureaucracy, "thrice removed from the Assistant Secretary's policy level."[7] However, the Eisenhower administration only agreed with a general policy goal of increasing tourism and, perhaps, some additional funding. It believed that the Commerce Department already had the general legal mandate to promote travel and did not need any further congressional authorizations. In particular, it opposed a semi-autonomous agency or one with its *own* offices abroad (US Senate 1960b, 16–19; US House 1960, 6–8). For a Democratic Congress, here was an opportunity to be more probusiness than a Republican president, especially in an election year. In 1960, a bill passed the Senate,[8] but not the House.

A few weeks after taking office, as part of a larger economic plan, President Kennedy endorsed legislation to promote foreign travel to the United States "along the lines" of the 1960 bill (Kennedy 1962, 63). Congress passed it with alacrity. On June 29, 1961, Kennedy signed the International Travel Act of 1961, creating in the Department of Commerce a United States Travel Service (USTS).[9] News coverage claimed, inaccurately, that USTS was the "first" of such a federal agency and program.[10] While intended to promote foreign tourism to the United States, the Department of Commerce stated that it was working with private industries "to promote travel to and *within* the United States" (1968, 7, emphasis added). In the context of the federal bureaucracy, USTS was a relatively modest operation, with headquarters in the Commerce building and seven overseas offices (8). Its funding from Congress started at $3 million in 1961 and gradually increased to $13.5 million in 1979 (Airey 1984, 273).[11] USTS existed throughout the 1960s and '70s, but with increasing complaints by the industry that it was not enough of a major effort by the federal government in terms of funding, staffing, and power.

There does not appear to be substantial academic literature judging and evaluating USTS as a federal agency and its record. In 1969, Thornton concluded that if judging USTS's record by increases in foreign tourism to the United States and reductions in the outflow of dollars, then the agency was not particularly successful (Thornton 1969). A few years earlier, Santoro was somewhat more positive about USTS's impact on America's balance of payments (1967).

President Johnson Resurrects the Park Service's Domestic Travel Promotion, 1965–1969

Even though Congress had defunded USTB in 1949 (chap. 8), the 1940 law giving Interior's NPS the statutory assignment to promote domestic travel (chap. 5) had stayed on the books and never been repealed. It was still valid, simply that Congress was not appropriating any money to fund the program nor was the executive branch requesting renewed financial support. By now, it had been nearly two decades since the closing of NPS's USTD. Nevertheless, even without an active travel promotion office, Interior was usually viewed as an integral part of any effort to promote foreign travel to the United States because national parks were world renowned and major destinations of tourists visiting America. Such parks as Yellowstone, the Grand Canyon, and Gettysburg attracted travelers from around the world. That meant any new and coordinated federal effort to promote foreigners to visit the United States ought to consider Interior and NPS, at least in policy planning.

Interior and NPS were itching to resume their vestigial role in travel promotion. A bureaucratic opportunity opened up in 1964 when Congress enacted a law calling for 1964–65 as "See the United States" years. The president was asked to promote this campaign "through such departments or agencies of the Federal Government as he deems appropriate."[12] This fit in perfectly with the mission the 1940 law assigned to Interior. The new NPS director, George B. Hartzog Jr. began lobbying the White House for a role in the See America campaign.[13] He also submitted to Interior Secretary Stewart Udall a plan for the president to designate Udall as head of any inter-departmental group assigned to implement the executive branch's participation in the See the USA campaign.[14]

Increasing domestic tourism also emerged as part of the policy and political response to the continuing (and growing) balance of payments deficit. Gradually, it became apparent that increasing foreign visitors to the United States was not the only way to improve the financial problems caused by American tourists going abroad. After all, every dollar that an *American* citizen spent on domestic tourism instead of foreign travel was one less dollar added to the trade deficit. Increasing domestic travel therefore emerged as part of the solution to the economic problem of the imbalance of payments. Furthermore, due to the peculiarities of geography, increasing domestic tourism was a major economic (and, hence, political) issue for Hawaii. At the time, most of its tourists came from the US mainland, so increases in promotion of domestic travel would be of direct economic benefit to the islands. This created a salient and ongoing interest in the subject of federal promotion of domestic travel by Hawaii's congressional delegation.

In 1965, President Johnson created a special cabinet task force on travel to deal with the balance-of-payments problem and appointed Vice President Hubert Humphrey to chair it. Somewhat oddly, its membership included Secretary Udall even though the department had no ongoing program to promote tourism, let alone foreign tourism. It was a quiet signal confirming that Udall's and Hartzog's efforts were having some success at getting Interior back in the travel game. In a speech to the National Association of Travel Organizations, Humphrey said that one goal of the effort was to "stimulate greater travel here by our own citizens." For example, he cited that Udall and NPS had already "established a special task force on Washington, D.C. The objective of this group is to make Washington, D.C., a pilot project for tourism."[15] NPS eventually developed a "travel information program" as a model cities template (US House 1970a, 100). Two years later, speaking to the nonprofit Discover America organization, the vice president reiterated the importance of increasing tourism "within this country," not just from abroad.[16]

In November 1967, in further pursuit of economic policies to deal with the trade deficit, Johnson appointed an Industry-Government Special Task Force on Travel. Again, indicating the results of internal lobbying, the membership of the task force included Harry M. Shooshan, Interior's deputy undersecretary for programs.[17] The group submitted its report to the president in mid-February 1968.[18] The report, as would be expected based on its mission, focused on increasing international tourism to the United States. However, which federal department should do so? Tiptoeing through the bureaucratic tulips, the task force was agnostic whether any new or reformulated administrative entity should be in "the Department of Commerce or be shifted to the Department of the Interior" or to LBJ's new kid on the cabinet block, the Department of Transportation (DOT) (Industry-Government Task Force 1968, 43).[19] Indicating the importance of this new department vis-à-vis tourism, in mid-1967, Representative Harley O. Staggers Sr., chair of the House Interstate and Foreign Commerce Committee, introduced a bill to transfer the responsibility for implementing the 1940 law from Interior to DOT.[20] While it did not pass, it was a vivid reminder of the latent power vested in that old law and how other bureaucracies coveted it.

One of the task force's specific recommendations almost off-handedly mentioned Interior's legacy statutory mandate for domestic tourism. The central focus of the recommendation was that travel brochures produced by various federal agencies should be translated into foreign languages for distribution abroad. Tacked on at the end of this suggestion was a different subject, only vaguely related to translating travel brochures: "The Interior Department should exercise the authority granted it by Congress in 1940 to engage in such promotion" (25). This was particularly odd because the subject of Interior's 1940 travel promotion mandate had not been mentioned at all in the text of the report preceding to this recommendation. It came out of the blue, almost as though it had been tacked on at the last minute.[21] Nonetheless, it was an open invitation to NPS to get back in business. The president quickly announced that the report would "receive prompt attention"

(Johnson 1970, 244). That was all Udall and NPS director Hartzog needed in terms of an administration endorsement and encouragement to act. They decided to revive a travel promotion program.

By now, it was late winter and early spring of 1968. FY 1968 was already more than half over, and Congress was beginning to delve deeply into President Johnson's budget proposal for FY 1969. At a Senate Appropriations Committee hearing in February 1968 for FY 1969, Hartzog talked about how national parks were major attractions, but that strict congressional limits on publication budgets hampered more extensive circulation of NPS brochures—thus reducing the potential of stimulating more tourism (US Senate 1968, 951). Similarly, a few weeks later, at a hearing of the House committee on the FY 1969 budget proposal, he also alluded to an NPS role in the Discover America campaign. Ever-so-vaguely, he also mentioned NPS's role in the campaign "to encourage foreign travel *as well as domestic travel* to our great places of history and scenery" (US House 1968, 404, emphasis added).

The Park Service then invoked a Capitol Hill procedure called reprogramming, which entailed shifting already-appropriated funds from one approved purpose in the agency to another. In April, after receiving permission to do so from BOB, it asked the two appropriations committees to permit a modest reprogramming to resume a small domestic travel promotion effort based on the 1940 law. NPS emphasized that the statute gave it the duty to promote all domestic travel, not just park-related visits. The reprogramming request was for $30,000 for the remainder of FY 1968 and $100,000 for FY 1969. (A reminder that the 1940 law [chap. 5] had authorized a maximum appropriation of $100,000 a year.) For FY 1968, the funding would cover two new permanent hires of travel liaison specialists (one at GS-15, the other GS-13), a secretary, a temporary travel liaison assistant (GS-9) to staff a welcome and information booth at JFK airport, and publication of posters. For FY 1969, the reprogramming would further cover three more staffers and increase the locations of welcome booths to airports in Washington, Chicago, and Honolulu.[22] The chairs of the two committees signed off on the reprogramming.[23] NPS was off to the races.

But, politically, time was short. On March 31, 1968, Johnson announced he would not run for reelection.[24] For Udall and NPS, the issue was how much they could accomplish during the remaining months of the lame-duck administration. After all, everything would change on January 20, 1969, when a new president would be sworn in. By then FY 1969 would be about half over, and the president's budget proposal for FY

1970 would have been submitted in early January. Therefore, they wanted to create a new status quo, a fait accompli, so that whatever they could institute would be harder to undo by whomever the next new president would be. Certainly, if Vice President Humphrey were to win, his support for continuing Interior's initiative was assured. But one of two other major Democratic candidates, Senators Robert Kennedy (D-NY) and Eugene McCarthy (D-MN), could be the party's nominee. Who knew what their policy positions would be? As for the Republican race, during the spring of 1968, Richard Nixon was gradually emerging as the leading candidate, but other serious candidates included California governor Ronald Reagan and New York governor Nelson Rockefeller.[25]

Udall and Hartzog wanted to create as many facts on the ground as possible. They sought to set up a bureaucratic infrastructure and momentum that might be able to withstand a change in administrations, even a change in the president's party. Using the reprogrammed funding for FY 1968 and 1969, they sought to accomplish four key elements of sustainable public administration: they hired a director, created a unit within NPS, recruited an advisory committee, and were funded in LBJ's last budget request (for FY 1970). However, before that, the first step entailed departmentwide planning. After all, even though NPS was the bureau with the assigned authority of the 1940 law, many other bureaus in the department had activities involving tourism and travel. They included the Bureau of Outdoor Recreation, the Bureau of Land Management, the Bureau of Indian Affairs, the Fish and Wildlife Service, and even the Bureau of Reclamation (because dams—especially the Hoover Dam—were major tourist attractions). In July, Secretary Udall set up an intradepartmental task force, chaired by Shooshan, to create a departmentwide plan that would be agreeable to all the bureaus—each jealously guarding its own turf.[26]

This triggered a major in-house fight. Shooshan and the other bureaus wanted to recommend a new departmentwide Office of Travel and to seek new comprehensive legislation to supersede the 1940 law. A first draft of the report depicted the 1940 law as antiquated and limited in scope, "only a footnote in history."[27] Hartzog hit the roof. It looked to him like the other bureaus were trying to horn in on NPS's already-existing statutory authority and obtain new legislation giving all of them a role in promoting travel. He particularly objected to the characterization of the 1940 law as outdated, citing the details of the statute as fully relevant to circa 1968 tourism, such as promoting domestic travel, displaying exhibits, disseminating information, and publishing brochures and materials.[28] Hartzog called Udall to complain, and that led Shooshan to back down.²⁹ The final version of the report was something of a bureaucratic compromise. It called for a departmental-level Office of Tourist Development, while retaining management and programming in the individual bureaus. It specifically recognized the continuing validity of the 1940 law because it "gives the Department the authority to respond immediately" without the need to wait for any new legislation or other external approvals.³⁰ The secretary promptly approved it.³¹

However, in the long run, there was a need at least to amend the current law because it had authorized a maximum of $100,000 for annual appropriations to NPS to fund it. That was now a very limited ceiling because, due to inflation and cost of living, $100,000 in 1940 only had the buying power of about $40,000 in 1968. Moving fast, NPS quickly arranged for friendly legislators to introduce in early October identical bills in the House and Senate to lift the funding cap from $100,000 to $3 million. The chief sponsor in the Senate, Henry "Scoop" Jackson (D-WA) said that the current cap "prevents the carrying out of a promotion program of sufficient scope to meet the national needs of today."³² It was very late in the session, and neither bill got a hearing (let alone passing), but by doing this, NPS had staked its claim on the subject and was setting the foundation for its consideration in the next Congress. By then it could be depicted as old business!

Udall was actively involved in accomplishing as much as possible for Interior's travel role before the end of LBJ's term. In late October, he addressed a conference of participants in the Discover America campaign and announced the department's initiative of increasing domestic tourism. Based on the number of annual visits to Interior's many sites, he said the Department had a long-time "historic role as 'host for the Nation.'" He emphasized that the department was supportive of Commerce's USTS and not trying to duplicate anything it did. One of the elements of Udall's initiative was that "the National Park Service established a Division of Tourism."³³ A few days later, Hartzog followed up with the announcement of the hiring of Ben Butterfield to be the director of the new NPS Office of Travel. Butterfield had worked in Commerce's USTS beginning in 1961. At the time, he was its marketing director. In announcing the appointment, Hartzog emphasized that the ongoing authorization for NPS to promote domestic tourism was based on the 1940 law and that the law assigned NPS responsibility for travel to all American destinations, not just national parks.³⁴ Finally, on November 4 (the day before the presidential election), Udall signed Secretary's Order 2912, assigning the responsibility for overseeing tourism promotion to the assistant secretary for Fish, Wildlife, Parks, and Marine Resources. The secretary also directed NPS to create as part of its senior management the position of "Assistant Director for Tourist Development."[35]

The haste with which all this was being done was reflected in the varying titles given to the effort. Udall's speech had announced a *Division of Tourism* in NPS, while Hartzog's appointment of Butterfield was as head of the *Office of Travel*. (A reminder that the original USTB name was changed in 1938 because the term travel was viewed as more encompassing than tourism [chap. 3].) Also, from the bureaucratic perspective, an office is generally of a lesser status than a division. Further confusing things, the Secretary's Order established a position of assistant NPS director for tourist *development*. It changed again a few months later. The name of the unit was the Office of Travel and Information Services, and Butterfield's title was now NPS "Assistant Director (Travel and Information Services)."[36] The confusion was further compounded in early 1969, when it was referred to in official correspondence and press coverage as the "travel division," the same title as the post–WWII effort (chap. 8).[37] In March, in congressional hearings on the FY 1970 budget, NPS director Hartzog referred to it variously as the Office of Travel and Information Services, the Travel Division, the Division of Travel, and the Information Division (US House 1969b, 967, 1034; US Senate 1969a, 1630).

The next step was to revive the travel advisory board, another component of the 1940 law. Again, Udall and Hartzog were in a hurry to set it up before the new president would come into office. A benefit to having the board in place before the January 20 inauguration would be the status of the public members. As leaders in the travel industry and of prominent standing, it would be awkward for the new administration to pull the rug out from under them. From a public relations point of view, this would *look bad* as well as generate ill feelings from those individuals and the organizations they were affiliated with. Udall and Hartzog knew it and wanted that additional chink in their armor to be in place when Nixon was sworn in. Based on recommendations from NPS, in November Udall signed letters inviting six men to serve on the board.[38] Most accepted. In mid-December, he announced the membership of the board, including a tour operator, an executive with Holiday Inns, representatives from the American Petroleum Institute and the National Association of Travel Organizations, a travel magazine executive, and the associate publisher of *Saturday Review*, a high-brow weekly. His announcement included the explanation that the revival of the board was based on the 1940 law, reflected "the recently expanded travel promotion effort" by NPS, and that the focus would be on all domestic travel, not just parks.[39] As usual, the Sunday travel sections of several major metropolitan papers covered the announcement.[40] Bureaucratic politics continued unabated regardless of the presidential interregnum. On December 18, outgoing Democratic Secretary of Commerce Cyrus Smith wrote Udall of his concern about the reactivation of Interior's travel promotion. He asked Udall not to take any "unilateral action."[41] The letter did not have any impact on Hartzog and Udall. For them, it was full steam ahead until noon on January 20, 1969.

The administration took two more actions to jump-start NPS's domestic travel promotion before President-elect Nixon would be inaugurated. On January 15, President Johnson sent his annual budget message to Congress, in this case for FY 1970. It is an oddity of timing in American government that the outgoing president submits a budget to the newly sworn-in Congress for a budget year that would not start until five months after leaving office. LBJ's budget proposal included increasing NPS funding "to expand the program to promote domestic travel" (US House 1969a, 605). The next day, Assistant Interior Secretary Max N. Edwards sent a letter to the House Speaker requesting legislation to amend the 1940 law by repealing the $100,000 cap on annual appropriations and replacing it with an authorization "for such sums as may be required," that is, no cap. Passage would then permit funding for FY 1970 for $225,000 that had been included in President Johnson's budget request the previous day (US House 1970a, 99–100).[42] The timing of the letter, sent to the new Congress but from the outgoing administration, permitted the request to be the pending and official executive-branch request unless and until the new administration might revoke it.

President Nixon's First Interior Secretary, Walter Hickel, 1969–70

President-elect Nixon nominated Alaska Governor Walter J. Hickel to be Secretary of Interior. Four days after Hickel took office, Hartzog notified Hickel that the first meeting of the travel advisory board was set for February 4 and invited him to attend.[43] Slyly, the agenda for the meeting included distribution of a new department-issued map of Alaska to be used to promote tourism there. (It had been prepared by a different Interior bureau, the US

Geological Survey.) Two weeks after the meeting, Butterfield's office released a new publication titled *It's Easier than You Think to Vacation in Alaska This Spring*.[44] Also the Bureau of Land Management (another Interior bureau), released a second edition of its brochure about public lands in Alaska called *Room to Roam*. This was all catnip for the new secretary. Demonstrating to his home state the benefits that would accrue to it because he was secretary was irresistible, especially if he had any ambition for another statewide election, such as running for the US Senate.

In February, the new secretary of commerce, Maurice Stans, picked up where his (Democratic) predecessor had left off and wrote Hickel of his concern about Interior's travel office. Diplomatically, he suggested "an acute need for close coordination" and suggested, "this situation warrants our personal attention and I would like to discuss it with you at the earliest opportunity."[45] Already, Hickel had "gone native," the political slang for a presidential appointee quickly taking on the agenda of the bureaucracy rather than of a White House–centric official. He was willing to defend his department's turf based on long-standing interdepartmental quarrels that had begun decades before he arrived. Giving Stans a bureaucratic cold shoulder, Hickel did not reply until two months later. He stated that "on the basis of my cursory review of the situation," NPS's promotion of domestic tourism was complementary, not duplicative, of USTS. He called for cooperation and coordination, but nothing more (US House 1973a, 20–21).

Public Relations

With the maximum appropriation of $100,000 a year that had been authorized by the 1940 law (and roughly half of that was salaries for six employees), the new travel office was able to mount only a relatively modest promotional program. During Hickel's tenure, the PR program focused mostly on free publicity, including public service announcements on radio and TV, prepackaged feature articles and photos distributed to newspapers, public service advertising for print media, and a monthly "Travel Filler" for use by magazines and newspapers for empty spaces in their news hole. Other promotional efforts were through coordinated campaigns with other federal agencies, state travel offices, and tourism-based businesses. In total, Hartzog estimated free media publicity valued at $1.25 million (US House 1970a, 112–14, 122).[46] By early 1971, the office had placed 796 items in all media and conservatively estimated that this PR outreach totaled 230 million impressions (i.e., audience members) (US Senate 1971, 2670). The office also produced a series of travel maps for sale through GPO and for distribution to travel agents (US House 1969b, 981, 1103).

Butterfield attended conferences to give speeches about the work of his office. In mid-1969, he linked the increased interest in environmental protection with NPS's work (Butterfield 1970a).[47] The next year, addressing a conference on travel research, he released new NPS data indicating trends in domestic travel statistics: the large proportion of travel was by families (rather than group tours) and that parks were a primary destination for most of those travelers (Butterfield 1970b). The dominant mantra and theme for vacations was now "getting away from it all."

Bureaucratic Politics and Power Struggles on Capitol Hill, 1969–70

During the presidential interregnum, Hartzog and Butterfield were electrified by an editorial in the newsmagazine for travel agents. It weighed in on the record of Commerce's USTS and lamented its stewardship of international travel promotion. USTS had "failed to live up to expectations." Now, "money and fresh leadership appear an urgent necessity." Noting Interior's mission of promoting domestic tourism, the editorial observed, "the Interior Department has better qualifications and credentials in the tourist field than the Commerce Department and its USTS satellite." Therefore, "considerable validity exists for transferring USTS to the Interior Department, thereby putting the Government's travel program under one roof."[48] As an expression of the views of travel professionals, the editorial had the potential of putting wind in the sails of Interior's efforts to reestablish a travel office and to deflate Commerce's counterpart ambitions. In the zero-sum trench warfare of bureaucratic turf battles, such an editorial might shift the balance of power to Interior.

After the new administration and Congress took office in January 1969, the turf war with Commerce's US Travel Service and congressional funding became inextricably linked. Hartzog and Butterfield's primary goals were to fend off efforts by USTS and its Capitol Hill allies to defund it or to force it to merge into USTS, while at the same time passing new legislation to lift the authorization cap of $100,000 embedded in the 1940 law in time for FY 1970 appropriations. The latter required intricate timing and political luck. For the two appropriations committees to be *able* to fund the travel office above $100,000 in FY 1970 required passage of authorization legislation. For that to happen, the two commerce committees would have to recommend a bill, for both houses to pass it, and for it to be signed by the president—all this before Interior's FY 1970 appropriation bill would be finalized.

It will be recalled (in the preceding subchapter) that in the waning days of the Johnson administration, the Interior Department had requested such legislation and proposed an unlimited authorization level. A few weeks into the new administration, Congressman Harley Staggers (D-WV), chair of the House Commerce Committee, wrote to ask Hickel if the department (and, impliedly, the administration) still supported that legislation? Staggers noted that the committee's general practice opposed open-ended authorizations. Given that, what authorization level did Hickel request for the next three fiscal years (US House 1970a, 101)? Hickel replied that he (and BOB) supported LBJ's legislative request for an open-ended cap and declined to provide specific authorization figures, reflecting the bureaucracy's continuing preference for no funding limitations.[49]

In the meantime, the appropriations committees could not wait. Based on the LBJ budget submission, the request for the travel bureau had been for an additional $125,000 for FY 1970, for a total appropriation of $225,000. However, by the time the House Appropriations Committee was ready to report the bill out, the standing committees had not acted on bills to increase the authorization beyond what was in the 1940 law. The maximum the committee could appropriate for FY 1970 was still $100,000, the amount that had been approved for FY 1969. Sending a strong signal of skepticism toward Interior's separate tourism office, the committee recommended only $45,000 for FY 1970 (US House 1969c, 24). The counterpart committee in the Senate was friendlier. Subcommittee chair Alan Bible (D-NV) suggested to Hartzog at the appropriations hearing that he wanted to see closer coordination between Interior and Commerce (US Senate 1969a, 1730). This indicated he understood the different missions of the two and recognized how much his home state benefited (like Hawaii) from domestic travel promotion. As a result, the subcommittee and committee recommended allocating $100,000 (US Senate 1969b, 20). In conference, the House receded, and the travel office, for the second year in a row, received an appropriation of $100,000 for the fiscal year (US House 1969d, 8; US House 1970b, 2).

Separately, in 1969–70, the two Commerce committees were dealing with multiple bills relating to tourism promotion, some friendly to Interior, some friendly to Commerce. Bills leaning to the latter would have ended Interior's travel mission, transferring all tourism promotion, domestic and international, to Commerce's USTS. Other bills entailed amending the 1940 law to an annual authorization higher than $100,000. The two behemoth departments were maneuvering on the Hill (and with their external constituencies) to vanquish the other's ambitions. Relations were strained and communication negligible. USTS largely ignored Hartzog and Butterfield's efforts to meet at all, let alone agreeing to a compromise nonaggression pact.[50]

In advance of a major House Commerce subcommittee hearing on the multiple pending bills, the (acting) secretary of interior revoked Hickel's 1969 letter requesting no funding limit and recommended instead specific authorization levels: $250,000 for FY 1971, $750,000 for FY 1972, and $1.5 million for FY 1973 (US House 1970a, 122). At the hearing, Congressman Spark Matsunaga (D-HI) testified with vehemence that his home state needed the federal government to promote domestic travel (as well as international) because so many tourists came from the mainland. Interior's program was vital to his state, he said (US House 1970a, 101–08).

After contentious lobbying and accusations, the House Commerce Committee decided on a political compromise and bureaucratic cease-fire. It recommended two separate bills. One would significantly expand the authorization for funding USTS for the next three years. The other authorized increasing the funding cap for NPS for only two years: $250,000 for FY 1971 and $750,000 for FY 1972 (US House 1970b). The reauthorization of USTS also included funding an independent National Tourism Resources Review Commission to examine federal promotion of international and domestic tourism and recommend future programming and funding levels (including financial grants to states). In particular, its assignment was to make a recommendation *where* such federal programming should be housed (US House 1970c). The implication was that, depending on the commission's report, after FY 1972, there might be no NPS domestic travel promotion.

Senator Daniel Inouye (D-HI) was a majority-party member of the Commerce Committee. He took the lead in Senate consideration of the NPS bill and was the named sponsor of the committee report (making him the presumed floor manager for debate on the floor of the Senate). While concurring in the House authorization levels, he made sure that the report stated an emphatic support for increasing domestic travel promotion. Approving this bill "would be another progressive step toward a more creative and expanded domestic travel promotion program. A vital, expanding domestic travel program is in the economic interest of the United States. It will also foster an understanding among our people of the great significance of their national and cultural heritage" (US Senate 1970, 3). By the time this bill reached the floor of the Senate, all the political controversies surrounding federal travel promotion and bureaucratic turf had been so well addressed in the temporary compromise (relating to a new commission to study the matter) that it passed without debate and by unanimous consent.[51] President Nixon signed it in December 1970.[52]

By then, however, there had been major political developments regarding Interior's leadership, and there were some rumors about NPS's as well. During 1969 and 1970, Hickel's service as secretary had been increasingly bumpy from the perspective of the White House. Initially, the issues related to the Interior's portfolio, with Hickel making decisions that tilted toward environmental protection over business interests. Then, in late April 1970, President Nixon announced that in order for the United States to prevail honorably in the Vietnam War, he was ordering an invasion of Cambodia. Campuses erupted with student protests, and, four days later, at Kent State University in Ohio, the state's National Guard shot at protesters and killed four students.[53] Hickel felt the White House was mishandling student dissent by ignoring and discrediting it. Two days after the shootings, he wrote a private letter to the president urging Nixon to change the stance "which appears to lack appropriate concern for the attitude of a great mass of Americans—our young people."[54] It quickly leaked. Nixon was furious, but felt any immediate reaction would further galvanize protests to his policies and perhaps even affect the results of the November elections for Senate and Congress. So he stewed privately and froze Hickel out. During that period, Hickel became more combative and outspoken politically.[55]

Three weeks after the election and the day before Thanksgiving, Nixon personally fired Hickel at a White House meeting.[56] Two days later, the Friday of the holiday weekend, White House aide Fred Malek went to the Interior building to clean house. He fired six senior officials.[57] Most of them were part of Hickel's personal staff, but the highest ranking person fired that day was a member of the subcabinet, Assistant Secretary Leslie Glasgow, who oversaw NPS and other related bureaus.[58] During the next workday, Monday November 30, rumors were rife that Hartzog and another official were about to be fired, too, because "both [were] career officials appointed to their present posts by former Interior Secretary Stewart Udall, a Democrat."[59] The White House denied any more personnel changes were pending. Instead, it would wait until Hickel's successor took office, and then he could decide on any further changes.

President Nixon's Second Interior Secretary, Rogers Morton, and the End of Interior's Domestic Travel Promotion, 1971–1973

The day he fired Hickel, President Nixon announced he was nominating Rogers Morton to replace him. Morton was a congressman (R-MD) and, simultaneously, chair of the Republican National Committee (RNC), having been selected when Nixon won the 1968 election. The latter position was a strong indication that Morton was a Nixon loyalist, a party man, and a team player. Morton was quickly confirmed by the Senate and took office in late January 1971.

Two months later, Nixon announced a new major initiative to reorganize the federal executive branch. Public administration reformers and presidents had been trying during most of the twentieth century to eliminate duplicative programs, move related programs in separate cabinet agencies into the same agency, and improve overall the management of the executive branch. In most cases, major reorganizations would require Congress to approve changes in the statutes, thus upsetting the turfs of committee jurisdictions. No chair wanted to lose power, nor did special interest groups benefiting from the status quo. Generally, both Capitol Hill and private-sector businesses opposed changes to "their" agencies. Therefore, most presidential grand reorganization efforts were stillborn (Lee 2010, 6–12). Nonetheless, Nixon tried. He called for a grand reorganization that would restructure the cabinet into coherent departments based on function. All like bureaus and programs would be consolidated under one roof. Nixon asked Congress to approve bills creating four superdepartments reflecting this functional principle: Community Development, Human Resources,[60] Natural Resources, and Economic Affairs. Most of Interior, including NPS, would be in the Department of Natural Resources, and most of Commerce would be in Economic Affairs. The president's plan for a Department of Economic Affairs explicitly mentioned that it would include USTS (US OMB 1971, 250).[61] NPS's travel office was not mentioned in the president's proposal, perhaps because it was too small to address; perhaps because everyone was waiting for the Tourism Review Commission's report. Given that the principle of reorganization to bring like activities under one roof, it was logical to assume that any fine print of Nixon's reorganization would include bringing domestic and overseas tourism promotion together.

A more explicit signal of the Nixon administration's lack of interest in NPS's travel program came two months before unveiling his grand reorganization plan. In January 1971, the president's budget proposal for

FY 1972 contained no funding request for Interior's travel office. He did, however, ask for funding for the Travel Review Commission. This money was needed for "determining policies and programs which will assure that domestic travel needs . . . are adequately and efficiently met" and if a single federal entity "should be designated to consolidate and coordinate tourism research, planning and development activities presently performed by different agencies" (US House 1971, 985). Therefore, this budget plan telegraphed where the administration stood on NPS's domestic travel promotion and the 1940 law. A few senators lightly pushed back to maintain some funding for Interior's travel office. One tried to make the case that the fully authorized amount of $750,000 should be funded in FY 1972. NPS carefully responded that this level of funding would permit the travel office to have eight permanent employees, a printing budget of $250,000 and an advertising budget of $360,000 (US Senate 1971, 2670, 3730). However, given the president's budget plan, NPS could not *request* it. And no senator vigorously followed up to pursue such an amendment to the Interior appropriation bill. Another negative signal was the new secretary's plan to reorganize the internal structure of the current department, partly as a prelude to Nixon's larger reorganization plan. Morton's plan included gaining stronger oversight of NPS, which was explicitly named in news coverage as one of the "various baronies" of the department that were too autonomous from secretarial control.[62]

A month after Morton took office and with reorganization in the air, Commerce Secretary Stans decided to try again in getting control over domestic travel promotion. The impact of the new regime at Interior quickly manifested itself. The ax fell on April 5, 1971. Morton replied to Stans's inquiry that, after reviewing the status quo, he "agree[d] that a consolidation of these two functions into one Department could promote their effectiveness" (US House 1973a, 21).[63] Reversing more than thirty years of history, Commerce had now won the turf war. Morton demonstrated that he would not "go native" as Hickel had and instead would pursue policies preferred by the White House over those of the bureaucracy. In particular, Morton did not feel he was obligated to continue protecting long-standing turf wars with other departments. The parochial interests of departmental bureaus were not automatically persuasive for him. Just because something was the way it was and alternatives were negative to the power of his department was not a compelling reason for him to go along with it. More important to him were larger principles of political harmony in the cabinet, being a team player, loyalty to President Nixon, and the organizational logic of consolidation. In particular, merging tourism promotion into one office was of a piece with the administration's larger effort at comprehensive reorganization of the executive branch by function. In a reverse twist of history, Morton's action was the mirror image of Ickes's success in 1939 at getting the commerce secretary to agree to Interior receiving the role of promoting domestic tourism (chap. 4).

However, as is often the case in government, this was easier said than done. Morton proposed to Stans that they use a president's reorganization powers to make the transfer. Beginning in 1932, Congress had almost continuously delegated to presidents the power to propose reorganizations of the executive branch, subject to congressional veto (Lee 2010, chap. 1).[64] The major limitation on reorganization powers was that a president could not propose anything that would conflict with a statute, such as tinkering with the legal authority for government to operate a particular program. When Nixon became president in early 1969, Congress relatively routinely extended to him reorganization powers through to April 1971.[65] In 1970, Nixon used those powers to create, for example, the Environmental Protection Agency (EPA), Office of Management and Budget (OMB), and the National Oceanic and Atmospheric Administration (NOAA) within the Commerce Department. In 1971, he proposed merging multiple volunteer agencies, such as the Peace Corps and VISTA into a new agency called Action. Congress approved all. More precisely, neither house vetoed those reorganization plans by passing a disapproval motion. Nixon's reorganization authority expired on April 1, 1971.[66] In preparation for that, Nixon requested that Congress renew his reorganization powers for another two years, through to April 1973.

Morton's April letter to Stan included a draft reorganization plan for the president to submit to Congress. The draft stated that "travel promotion activities of the Government can best be focused and coordinated by a single agency, and that agency should be the one with the broadest experience in the particular program activities of this kind" (US House 1973a, 22). The draft also stated that no budgetary or personnel transfers from NPS to Commerce would occur. Morton's assumption was that Congress would routinely extend Nixon's reorganization powers and that the White House would promptly submit the plan. He envisioned that the change would occur on July 1, 1971 unless congressional consideration went beyond that date. For reasons that are unclear, Congress did not approve the legislation extending Nixon's reorganization powers until December 1971.[67] That meant presidential authority to submit reorganization plans had lapsed from April to

December 1971. Therefore, Morton's plan had to be held in abeyance until whenever the renewal would occur. However, even then, Nixon did not submit the reorganization plan and did not do so through to their re-expiration in April 1973. In fact, the final report of the National Tourism Resources Review Commission in June 1973 stated, quizzically, that notwithstanding Secretary Morton's support, "this transfer had not been accomplished by 1972" (US National Tourism Resources Review Commission 1973, 5).[68]

In the meantime, at the end of FY 1971, the last funds appropriated by Congress to the NPS travel office had been expended, and no funding was requested by the administration nor initiated by Congress for FY 1972. A congressional report in November 1973 stated that "the program is now dormant" (US House 1973c, 3). In other words, this was similar to the status of USTB in 1943–46 and USTD from 1949 to 1968. The 1940 law giving Interior the exclusive responsibility for domestic travel promotion was still in effect, but was not funded. One of Butterfield's last official actions as head of the travel office was a September 1971 letter to USTS about cooperating on an effort to familiarize travel agents with some national parks.[69] Then the office quietly shut down. He moved on to a different career at NPS.

From December 1971 to April 1973, President Nixon never submitted Morton's reorganization proposal to Congress. However, during the winter of 1972–73, a much more significant event indicated Nixon's hostility to the bureaucracy in general and NPS in particular. After winning reelection in November 1972, he recommitted to getting control of the bureaucracy. One of his initiatives was to implement his major reorganization of the executive branch, but this time finessing congressional approval. Instead, he named three sitting or incoming cabinet secretaries to serve simultaneously as White House–based supersecretaries to manage the various departments and agencies in a way that reflected his earlier grand reorganization request to Congress (Lee 2010).[70] After the election, Nixon also asked all cabinet secretaries to offer pro forma resignations. He accepted many and moved some to other departments.

Morton was one of the handful of secretaries the president retained (at all) and kept in the same place. The day after Nixon announced Morton's reappointment, Morton fired Hartzog.[71] Nixonites viewed Hartzog hostilely, in part because he came up through civil service ranks at NPS, and therefore they assumed his loyalty was to "his" bureaucracy. Furthermore, because he had had been appointed by Democratic Secretary Udall, his political allegiance to the Nixon administration that much more suspect. In parallel to the secretary-level housecleaning and reorganization, Nixon was determined to obtain further control over the bureaucracy during his second term by dispersing White House staffers from his first term to subcabinet positions (i.e., assistant secretaries) and the next level down, bureau chiefs, throughout the executive branch (Nathan 1975, 67–68). This second-term strategy "moved many second-level White House aides to key management posts" (Nathan 1986, 51). Their loyalty to him would assure that those who ran the bureaucracy on a day-to-day basis would be *presidential* advocates rather than civil servants and defenders of bureau autonomy.[72] On December 13, 1972, Nixon met with and then announced the appointment of Ronald H. Walker, head of the White House advance office, as the new NPS director.[73] That Hartzog's successor was a political appointee with no significant background in parks was a stark signal of Nixon's hostility to NPS's autonomy and his determination to subordinate it to his priorities and agenda.[74] According to Nathan, of the many White House aides Nixon scattered throughout the executive branch, Walker's appointment was one that particularly occurred "amid controversy" (1975, 68).[75]

With all these changes, what of NPS's travel office? Due to the continuing White House passivity on any formal action regarding domestic travel promotion between 1971 and early 1973, congressional action forced the subject. In February 1973, Congressman Matsunaga introduced a bill to transfer domestic travel promotion authority from Interior to Commerce.[76] Later that year, he explained his rationale: "Such a consolidation . . . is essential to the success for any travel program, as we have learned in my home State of Hawaii."[77] Senator Inouye concurred.[78] This meant that Hawaii's congressional delegation no longer viewed the separate existence of a domestic travel office in Interior as necessary for its interests. The political tide had turned. This was reflected in a column in the *New York Times*'s Sunday travel section in April 1973. Based on the new political consensus that USTS was, of course, the proper location of domestic travel promotion, it complained that "Congress has refused to let it [USTS] *expand* to promote travel within America by Americans." As a result, "travel promotion in the United States is left to the private sector."[79]

Later in 1973, an omnibus travel bill reauthorizing USTS included a section transferring the authority of the 1940 statute to Commerce (US House 1973c). Given that Interior by now had no funding for this function, the statutory change did not entail any transfers of appropriations or personnel. President Nixon signed the bill into law on December 19, 1973.[80] So ended Ickes's fervent efforts in the late 1930s for Interior to have the portfolio of domestic travel promotion.

Afterword

President Carter was on the other side of the perennial debate about the appropriateness of a federal role in travel promotion. In 1979, he proposed wholly defunding USTS (Gunn 1983, 33). Congress rejected the suggestion. The next year, Congress tried to supersede USTS with a freestanding entity in the executive branch devoted exclusively to tourism and travel (US House 1980). President Carter vetoed the bill. In his veto message he criticized trying to create an autonomous US Travel and Tourism Administration outside the Commerce Department. He also objected to the new agency overseen by an advisory board with a majority of members from the private sector and opposed the agency submitting its annual budget request directly to Congress, hence beyond the purview of a *president's* annual budget plan. The agency "would be more responsive to special industry interests than the need for a coordinated Federal approach that will balance the needs of tourism against other national priorities," he said (Carter 1982, 2838).

President Reagan's first budget proposal also sought to eliminate USTS, while maintaining some funding for its programs and activities (Airey 1984, 274). When Congress considered passing a somewhat softer version of what Carter had vetoed, the conventional wisdom was that Reagan's philosophical opposition to big government and federal spending would trigger a veto.[81] Yet, surprisingly, in October 1981, Reagan signed the National Tourism Policy Act.[82] The new law abolished USTS and replaced it with the US Travel and Tourism Administration in the Commerce Department, headed by an undersecretary (Edgell 1983; 1984). Four years later, Reagan's budget proposal included abolishing the Travel and Tourism Administration as a way to cut federal spending.[83] As a compromise, Congress saved the office by funding it with a one-dollar fee for all travelers entering the US.[84] But that only lasted until 1996. During President Clinton's administration, the National Tourism Policy Act was largely revoked, leaving in the executive branch only a Tourism Policy *Council* in the Commerce Department (Edgell and Swanson 2019, 92, 192). A related nongovernmental National Tourism Organization died on the vine, largely unfunded.

But the industry and state and local governments dependent on tourism continued agitating for a federal role—particularly funding. That, in turn, led to the formation of a Senate Travel Caucus and a Congressional Travel & Tourism Caucus. This meant there was an in-house organized effort for increasing the federal role. In 2010, Congress passed and President Obama signed the Travel Promotion Act.[85] It created a very small Office of Travel

Promotion in Commerce and provided user-based funding for a public-private nonprofit called Brand USA. The federal revenues came from a surcharge on electronic visa issuances along with a requirement of matching funds from the industry. In 2020, when the authorization for the program was about to expire, Congress extended it until 2027 and increased the user fee from $10 to $17 to generate more revenue for Brand USA.[86] Commerce's renamed National Travel and Tourism Office continued limping along as an extremely minor federal entity, listing only thirteen professional staffers in 2020, barely a speck in the federal leviathan.[87] President Trump's proposed federal budget for FY 2021 recommended that Congress eliminate all funding for Brand USA for reasons of "fiscal responsibility and to redefine the proper role of the Federal Government."[88] But Congress was uninterested in such an abstract ideological argument and instead maintained the existing structure and program. Meanwhile, an independent academic study of Brand USA's record concluded tentatively that the organization's marketing had little effect on inbound international travel (Zavattaro and Fay 2019). These were merely the latest rounds of the apparently never-ending debate over the normative role of the national government in travel promotion.

Conclusion

The preceding chapters presented a full-length biography of a small and largely forgotten federal agency, the US Travel Bureau, from 1937 to 1973. It began as an administrative creation of Interior Secretary Ickes. Using funding from WPA and other economic emergency appropriations, it began operating immediately to promote tourism. Its existence was greatly enhanced when the 1939–40 Congress enacted a law formalizing USTB's status and funding within the Interior Department. Now USTB became a robust and vigorous agency—briefly. After the United States entered WWII, the bureau had an on-and-off history, sometimes functioning and sometimes a ghost agency. It largely suspended operations during WWII (though nominally in existence through February 1943), was revived after the war, suspended again in 1949, revived in 1968, defunded in 1971, and extinguished statutorily in 1973.

Paralleling its public administration profile, USTB also encompassed a complete cycle of the origins, gestation, life, and death of a federal law. After considering bills on the subject as early as 1930, a 1940 act declared that domestic travel promotion was an appropriate activity for the executive branch, that it should be funded by Congress, and that it should be in the Department of Interior. Notwithstanding the ups and downs of USTB as a funded and operating entity, this law lasted from 1940 to 1973. During that time, the law was amended only once. In 1970, Congress passed and President Nixon signed a bill increasing the maximum authorized funding level of USTB from the 1940 level of $100,000 a year to $750,000 for FY 1972. Then, in late 1973, Nixon signed a bill repealing that law and USTB ceased to exist.

Summary of Disciplinary Contributions

The Travel Bureau (and its renamed successors) was the first organized effort in the federal government to promote travel. Thus, its story adds to the literature and expands the knowledge base in several academic disciplines, including public administration, political science, American history, public relations, African American studies, and tourism and leisure studies. From the perspective of public administration, this recounting covered a complete case study of a bureaucratic silo in the federal executive branch, a birth-to-death narrative. In the context of the scale of federal agencies and federal spending, USTB was relatively small. Nonetheless, its existence had previously been unexplored in public administration history as well as in the history of the federal government. As such, the inquiry adds to this general literature.

Confusingly, the public administration term "bureau" in its title was something of a misnomer. In the general usage by Congress and the federal executive branch, the hierarchical category of bureau applied to very large agencies, usually the first-level subunits comprising a cabinet department and were directly under the secretary and relevant assistant secretary (US Bureau of the Budget 1942). Some of the most well-known and traditional such entities included the Census Bureau in the Commerce Department, the Bureau of Reclamation in the Interior Department, and the Bureaus of Internal Revenue and Engraving in the Treasury Department. These were very large administrative agencies, usually funded by annual budgets in the millions and staffed by thousands of civil servants. Other somewhat smaller, but still significant, bureaus included the Bureau of Labor Statistics in the Labor Department and the Bureau of Agricultural Economics in USDA. Compared to these longtime executive branch bureaus, USTB was a very small unit with a maximum appropriation of $100,000 (until its last years) and usually with less than a hundred staffers.

Compounding the confusion was its bureaucratic perch. With the exception of about a year before the passage of the 1940 law, USTB and its postwar successors were always a subunit within the National Park Service. This confusing terminology of a bureau *within* the bureau-level Park Service caused no end of misunderstanding. Besides violating the general federal nomenclature and usage, the 1940 law dictated that USTB (and its successors) had to be in NPS. This, in turn, triggered further and recurring misimpressions about its mission. At numerous congressional hearings, committee members assumed that the goal of the unit was to promote visits to the national parks. Repeatedly, witnesses for USTB and NPS had to re-educate committee members about the oddity that its statutory mission was to promote domestic tourism to all destinations, not just to national parks, not even limited to other Interior Department entities. They emphasized, often unsuccessfully as far as having their listeners absorb the message, that

USTB sought to promote all domestic tourism regardless of the affiliation of the destination, such as National Forests (run by USDA), state parks, and for-profit travel venues.

In particular, USTB was an unusual phenomenon in public administration because its main output was not a traditional governmental product or a tangible service. Similarly, its focus was not on serving a well-defined client, demographic, or constituency group. Instead, it was oriented to reaching the public-at-large. Its goal was to engage in external communications to *persuade* American citizens that they should travel more around the country. It was a peculiar manifestation of the role of public relations in public administration. This was public relations *as* public administration. More oddly, it was public administration *as* public relations.

What is especially surprising is the pass that Congress gave USTB on this issue of vigorous public relations. Plainly, USTB was a propaganda agency. It used all the tools of PR to influence the behavior of Americans, to persuade them to travel more. This was much more than neutral information or civic-minded public reporting. It could have been characterized by opponents as overt agitprop and brainwashing. Generally, Congress opposed bureaucracies having external communications programs. Agencies should be seen, not heard. This philosophical orientation became much more pointed with the rise of the New Deal. The conservative coalition on Capitol Hill vociferously objected to FDR's expansion of PR offices in executive branch agencies. This was propaganda, they said, for FDR, the Democratic Party, and Big Government. In particular, their criticism of federal PR was often aimed at Ickes and the Interior Department. Congress objected to him creating a department-wide Division of Information (to parallel USDA's), his construction of a radio station on the top floor of the new Interior building, and a radio version of his annual report that was aimed at reaching a mass audience instead a print annual report that no one read (Lee 2011, chap. 7). Perhaps the lack of legislative objection to USTB's propaganda was due to business's support for federal travel promotion. This endorsement from the industry was like a Good Housekeeping Seal of Approval on an otherwise objectionable activity by a bureaucracy. Or perhaps the silence was because love of country and getting to know it through travel transcended partisan and ideological differences. For whatever reason, a surprising aspect of USTB's history was the absence of controversy over its propaganda.

Bureaucratic politics and turf battles were a permanent feature of USTB's existence, as they commonly were in the federal executive branch (Wilson 2000, chap. 10). The Commerce Department asserted that any travel promotion, whether to encourage foreigners to tour the United States or to promote domestic tourism, was inherently part of its brief. Interior Secretary Harold Ickes aggressively countered that claim. He argued that the scope of his department was a better fit for domestic travel promotion because Interior encompassed more than the attractions of national parks and national monuments. Other departmental bureaus also had travel-related activities. These included outdoor recreation on lands of predecessor agencies of the Bureau of Land Management, wildlife refuges of the Fish and Wildlife Service, Indian reservations, US territories and possessions, and the monumental engineering structures of the Bureau of Reclamation, such as dams. (The Hoover/Boulder Dam was a major tourist attraction.) Stretching a bit, another unit of the department, the US Geological Survey, produced inexpensive maps that were frequently useful to the public-at-large when traveling (although these maps were not prepared and issued for that specific use). Ickes persistently pursued Commerce Secretary Harry Hopkins until the ailing secretary agreed to sign a letter to Congress supporting Ickes's claim.

Once in place, the law was a kind of bureaucratic gravitational field. The power of public law meant it could not be disregarded nor undermined. It was like the Rock of Gibraltar in the political and bureaucratic landscape. The only way to undo the policy determination by Congress would be by convincing Congress to change its mind and supersede the 1940 law with another authorization bill. (An appropriation bill could not do that.) In Washington, this meant a very steep climb. Ickes understood how hard it was to change the status quo of government. Indeed, his win extended far beyond his service as secretary. It stayed on the books for more than three decades. Yet the win continued to rankle the civil servants of the Commerce Department. Finally, in 1971, Commerce Secretary Maurice Stans convinced Interior Secretary Rogers Morton to reverse it and to agree to transfer the portfolio to Commerce. Even then, it took two and a half years for Congress to codify that agreement and for the president to agree as well.

For political science, the story of USTB also presented a relevance beyond the field's subcategories of public administration, FDR's presidency, and political history. For example, it also was a narrative of internal congressional decision making as well as the story of an external special interest lobbying for a program to benefit its parochial economic self-interest to be paid for by the taxpayers. Furthermore, USTB's history comprised a study of public policy from start to finish. In some respects, its demise is an example of a failed alternative, a focus that has often been absent in policy history. Regarding the academic discipline of public relations and mass communication, the field had originally focused largely on the practice in the business sector and then, to a lesser extent, on the nonprofit/NGO sector. USTB is an example of the rubric of PR practices in the public sector. This sector had been largely under-reported in the academic literature of this discipline. As such, additional knowledge of USTB as a federal PR agency contributed to a more textured knowledge base about the practice of public relations in the government.

The fields of leisure studies, tourism, travel, and recreation have emerged as relatively new academic disciplines. Extant literature in this field has often referred to USTB as part of the early history of a US governmental role in promoting travel and tourism. However, these references were sometimes only in passing or contained significant errors. Therefore, this study of USTB as the first federal agency to promote travel deepened scholarly familiarity with it, expanded the field's historical subliterature, and corrected previously published mistakes and misstatements. Regarding environmental studies, USTB was something of a precursor to the rise of ecology tourism. It recognized early in its existence, and increasingly after WWII, that conservation, pollution prevention, and protection of the environment could be motivators for a new and emerging form of American travel. (For a discussion of the significance of USTB to the academic discipline of African American studies, see next subchapter.)

Examining USTB's Historical Significance

Beyond the value that this historical report can contribute to several scholarly disciplines, USTB is particularly significant historically for three distinct and unusual dimensions of its existence. First, the Bureau quietly played a role in bestowing several political and policy benefits on the New Deal and FDR's presidency. Second, USTB disregarded the pervasiveness of racial segregation that was the status quo of the times (not just in the South). USTB went out of its way to serve African Americans and did so in a public and open way. Third, USTB reflected the ongoing normative and ideological arguments over the proper role of the federal government, in this case regarding tourism and travel.

USTB's Contributions to FDR's Political and Policy Goals

Notwithstanding its modest size, USTB fulfilled several disparate political and policy goals of the New Deal and the administration. These included economic stimulus, conservation, hemispheric solidarity, an antidote to travel limitations caused by the war abroad (before Pearl Harbor), general patriotism in the run-up to the US participation in the war, and—perhaps the most peculiar—enthusiastic business support for this New Deal program.

USTB's most obvious and publicly stated rationale related to the economic crisis of the Great Depression. Given the deeply embedded American cultural value of wanderlust, tourism was a method of increasing economic activity and (re)creating jobs. With the bureau coming into existence in 1937, it was relatively late in the Depression and the first New Deal. However, economic and employment problems were still pressing in the first years of FDR's second term. Sometimes forgotten is that the country experienced an economic downturn in 1938, virtually a resumption of the Depression. FDR suddenly reversed his push for a balanced budget, requested new emergency spending from Congress, and revitalized the National Emergency Council, his New Deal bureaucratic infrastructure to coordinate economic recovery (Lee 2005, 30). USTB's work during that period was very much in the context of promoting economic recovery by increasing consumption and individual spending, in this case on travel and leisure.

Before USTB came into existence, little quality research existed about the role of travel in the national economy. However, Washington's policy makers intuitively understood the economic impact of tourism. Practically every US senator and member of Congress had constituents who worked at hotels, summer resorts, ski resorts, parks, campsites, dude ranches, hunting and fishing lodges, travel terminals, or historical sites and for sightseeing, leisure cruise, recreation, or transportation companies. These politicians knew that increased travel would be beneficial to their voters. From the president on down, a new and coordinated federal effort to increase travel could be a novel and additional program to a passel of governmental initiatives to reverse the economic stagnation of the 1930s. USTB's later research concluded that the travel sector could be as big as the third largest component of the national economy. For policy leaders, the Travel Bureau's work would be indirect economic stimulus.

The popular narrative of FDR's legacy before Pearl Harbor understandably focuses on economic recovery. However, often lost in this abridged version of national history is how dedicated he was throughout this period to conservation of nature and restoration of lands despoiled by industry. Brinkley's (2016) detailed examination of Roosevelt's record in this area recounts the president's unprecedented, persistent, and continuous actions. He expanded the national park system, added national monuments, built infrastructure and facilities to promote outdoor activities, expanded wildlife refuges, improved land conservation, planted trees, and protected endangered species. FDR cannily understand that to protect his accomplishments from being rolled back he needed broad public support. Travel and tourism were ways to build this. He presumed, probably correctly, that new visitors to national parks would likely become adherents to the mission of protecting the country's beauty, vistas, and unique natural attractions.

The rise of Nazi Germany also had implications for the United States vis-à-vis the Western Hemisphere. German representatives were active throughout Latin America trying to increase disaffection with the US and to view Germany as a potent ally instead. To counter this, FDR and the State Department pursued many initiatives to promote closer ties between the United States and Latin America. One of these efforts was to promote more tourism by US citizens to Central and South America. Tourism in the other direction, of increased visits by Latin Americans to the United States, was viewed as equally important. USTB was involved in this hemispheric tourism promotion, even though it violated, strictly speaking, its statutory mission of domestic tourism. It received special appropriations from Congress to translate its brochures into Spanish and Portuguese and to create movie shorts and other promotions that could be circulated throughout Latin America.

European grand tours became increasingly common for Americans during the initial decades of the twentieth century. Mass travel across the Atlantic by large steamships offered a relatively inexpensive and reliable way to travel. However, European travel slowly became increasingly chancy even in the years before the beginning of WWII. For several years before Hitler's invasion of Poland in September 1939, there were tumultuous events that made Americans hesitant to continue traveling to Europe, and they wondered how safe it was. These unsettling developments included Mussolini's war rhetoric and eventual invasion of Ethiopia (October 1935), Hitler's remilitarization of the Rhineland (March 1936), the beginning of the Spanish Civil War (July 1936), Hitler's takeover of Austria (March 1938), and his threats to Czechoslovakia leading up to the Munich Accords (September 1938). With many popular European destinations for American tourists looking somewhat risky, USTB quickly pivoted to take advantage of the opportunity to increase domestic travel. Here was a solid and attractive substitute for international tourism and it was safe, to boot. (Besides, the natives spoke English.) Domestic tourism provided a kind of safety valve to maintain economic and political stability in the United States during these scary European events. It contributed to retaining a sense of normalcy in daily life notwithstanding international developments. Informally, USTB was pitching to these erstwhile and disappointed travelers to Europe, "You'll find equally fascinating and fun places to visit in your own country's backyard. Come see us!"

Another political and policy purpose that USTB contributed to before Pearl Harbor was to promote patriotism and love of country during FDR's gradual national defense mobilization. It began with his declaration of a limited national emergency in September 1939. Pearl Harbor was two and a quarter years away. Increases in domestic tourism could enhance civic pride and, consequently, support for policies to promote a strong national defense. True, domestic tourism could have had the opposite effect of supporting isolationism. However, for example, the army did not think so. Its extensive recreational activities for draftees often included visits to national parks, monuments, and other major patriotic sites. Some draftees had traveled little while growing up during the Great Depression. Their knowledge of the country was often quite limited. For them, these army-sponsored tours expanded their familiarity with other parts of the country and likely enhanced their understanding of importance of the national defense mobilization.

Perhaps the most unusual policy and political aspect of USTB was the support it engendered from the private sector. From the beginning of the New Deal, FDR's alphabet agencies and expansion of the role of the federal government in the economy faced unrelenting attacks by conservatives, Republicans, and business. They depicted the New Deal and FDR as enemies of private enterprise, antibusiness, covertly socialist, seeking to replace the market-oriented economy with Big Government, and that it was run by powerful and unaccountable bureaucrats and so-called nonpartisan experts. The administration, they said, was a threat to Americanism. Roosevelt was even sometimes accused of wanting to destroy the republic, to become an unelected dictator, and to have his sons be his successors.

For the wily Roosevelt, federal promotion of tourism was a kind of political wedge issue. It effectively peeled off a portion of the business sector from a generalized opposition to the New Deal and big government. Instead, they expressed vocal support for this particular New Deal program and bureaucracy. In the face of unrelenting hostility in general, business and chambers of commerce enthusiastically supported USTB. They lobbied Congress to pass a bill creating USTB and they lobbied to keep it alive and increase its funding. They were not concerned about the philosophical and abstract arguments relating to the proper role of the federal government. As far as they were concerned, if it was good for them, it was good for the country. Politically, for FDR this was a breakthrough and a crack in the monolithic political hostility to his programs. Some business liked some of the New Deal! They were saying nice things about something his administration was doing. Initially, the support came from businesses in states with major tourism business and, on their behalf, local chambers of commerce. However, the economic benefits of travel turned out to be widespread, with almost every state having some tourism-based activities. Eventually, one of the most conservative of the national business organizations, the United States Chamber of Commerce, publicly advocated for USTB's agenda and funding.

However, a fair question is who was coopting whom? In his classic study of cooptation (or, using the more modern term, capture theory), Selznick concluded that the Tennessee Valley Authority was coopted and captured by important local elites rather than the other way around (1984). Was FDR coopting the travel industry or was the travel industry colonizing the federal government to promote its narrow economic self-interest? After all, the federal executive branch was littered with agencies that catered to and were captured by private special interests (Wilson 2000, chap. 5). Organized economic interests wanted to have their "own" agency: farmers had USDA, unions had the Labor Department, banks had the Federal Reserve, ranchers had (what came to be called) the Grazing Service in Interior, and railroads had the Interstate Commerce Commission.

The same occurred with USTB. The travel business now had its own little federal agency to be its voice and to serve its economic interests and needs. When Congress approved legislation and authorized the agency, the statute explicitly directed the interior secretary to create an advisory committee and specified that its members should include "representatives of the various sections of the Nation, including transportation and accommodation agencies."[1]

Indeed, a year later, in the next round of the congressional appropriations process (for FY 1942), the industry vehemently lobbied the Senate to continue funding the bureau, even though the House had recommended zeroing it out due to the national emergency (chap. 6). Representatives of the American Hotel Association stated that USTB was already tangibly increasing the business of its members. Another said that, more generally, the bureau's record of accomplishment "has met with the widest approval" of the industry at large (US Senate 1941a, 219–20). The spokesperson for the National Bus Traffic Association testified on behalf of his members that the bureau "has done an excellent job" (222). A concrete example of such cooptation by organized interests of a government agency also occurred that year. Unrelated to the bureau, the industry had initiated a campaign for legislation to shift the observance of holidays to Mondays as a way to increase travel. Previously silent on the subject, in reaction to the desires of its business constituency, USTB then also took up this policy goal. It publicized and promoted the effort in several issues of its *Official Bulletin* (chap. 6). Here was a synecdoche of the broader phenomenon "that many public agencies represent and promote the interests of (mostly) well-organized groups" (Cook 2014, 211).

Cooptation is probably best understood as a two-way street, with each side gaining some benefit because, otherwise, the relationship would likely not be viable for any extensive time. A fair conclusion seems to be that FDR and the travel business coopted each other in a win-win relationship. Each got something they wanted. For the politically shrewd FDR, this transaction was more than good enough.

In summary, USTB may have been a relatively modest operation in the context of the federal government, but it nonetheless made specific and discernable contributions to the policy goals of the New Deal and to the political interests of Franklin Roosevelt's presidency.

USTB's Active Support for African American Travel

President Roosevelt was distinctly unenthusiastic about taking overt actions that would be seen politically as promoting equal rights for African Americans. The overcharged and emotional attitudes of many whites supporting segregation usually reflected their view that race was a zero-sum power dynamic. Gains for blacks meant an equal (or larger) loss for whites. FDR's timidity about taking initiatives in support of civil rights partly reflected his pragmatism in the face of the political clout of southerners in Congress. They could stymie so many things he wanted to accomplish. Therefore, the idealistic goal of fairer treatment of African Americans in American society—which he supported as an abstract principle—was one that he expediently demoted in importance relative to other policy goals. As something of a counterweight to FDR's posture, Mrs. Roosevelt was actively involved in efforts to reduce discrimination against African Americans and increasing their economic and educational opportunities, as was Ickes.

One of the most prominent examples of FDR's passivity on equal rights occurred in the spring and summer of 1941. During 1940–41, the vast increase in expenditures for arms production included very large contracts with many private-sector manufacturers. In turn, they sought to hire thousands of new employees. Maintaining the racial status quo, employers generally hired only whites for these jobs. When they did hire African Americans, it was almost exclusively for unskilled and lower-paid positions and most definitely did not involve placing an African American as a supervisor over white workers. Plainly, federal contracts were funding this discrimination. The private employment decisions of the private sector were one thing, but using federal tax dollars to maintain discriminatory hiring was another. The major national civil rights groups announced that they would be holding a March on Washington in the summer of 1941 to demand that federal arms contracts require equal employment policies. FDR was adamantly opposed to the March and realized that politically he was in a box. If he took any actions after the march occurred, Southerners and other racists would accuse him of caving in to black political pressure. If he took no action, then the broad voting support he traditionally received from black voters could dissipate. Eventually, a behind-the-scenes political compromise evolved. The African American leadership would call off the march. Then FDR would issue an executive order requiring that federal arms contracts would prohibit employers from engaging in discriminatory hiring decisions when spending those federal funds (Lee 2018, 97–98). The President followed it up that fall, just after Labor Day, when he issued a public letter to all executive branch department and agency heads reminding them that hiring decisions for the federal civil service should avoid discriminatory practices, too (Lee 2016, 95). Neither of these actions required congressional approval. In each case, FDR accomplished them by the stroke of a pen.

In the context of FDR's reluctance to take overt actions seen as favoring African Americans (at the supposed expense of whites), USTB's activist approach to African American travel vividly stands out as unusual. It explicitly sought to encourage blacks to travel and to enjoy the pleasures of tourism as much as whites did. The bureau's record on equal rights was a relatively quiet way for the administration to show some support for African Americans, but with lower visibility and therefore less likely to trigger a political price to pay with Southerners in Congress.

As recounted in chapters 3–6, USTB hired three African Americans to staff its Division of Negro Activities in the New York office. Charles McDowell and his two female assistants engaged in research about African American travel, issued directories of hotels and guesthouses that welcomed African Americans, and cooperated with private publisher Victor Green in compiling information for two of the annual editions of his *Negro Motorist Green Book*. However, USTB did not try to do this invisibly or secretly. It was public and visible to all. For example, McDowell regularly contributed a column on travel by African Americans in the official USTB newsletter that the New York office published and distributed widely in the eastern half of the country. (The San Francisco office published a counterpart for the western half of the US.) In another case, an issue of USTB's *Official Bulletin*, which had national circulation, included an article on its recently released publication aimed at African American travelers.

McDowell ceased working for USTB in late 1941. The archival record did not provide an explanation for this. However, documentation of the timing of his departure can pinpoint that it happened before Pearl Harbor. Therefore, the declaration of war and the impact it would to have on USTB programming was not the reason for his exit. While purely speculative, the most likely explanation is the gradual tightening of USTB's budget and the need to scale back and trim some of its activities. There is also the possibility that turnover in USTB's senior management might have put in place some men who were less enthusiastic about serving African Americans than their predecessors. However, this is purely a hypothetical speculation and should not be viewed as having any underlying historical documentation. Furthermore, Secretary Ickes was a strong supporter of civil rights and his role overseeing USTB did not change from the time of its establishment to after McDowell departed from the bureau.

It is important not to overstate USTB's record in this area. It did not overtly challenge the Jim Crow regime in the South nor its sub rosa existence in the rest of the country. Similarly, it did not get into public fights on the subject nor engage in explicitly provocative actions. It was not taking on the entrenched segregation of its time and trying to nullify it. It had little choice but to accept the-then legal principle of "separate but equal," with the "equal" part of the doctrine—as practiced by American governments—rarely the case. Rather, it tried to maximize travel opportunities for African Americans within the context of the status quo. Nonetheless, given the mores of the times and FDR's hesitancy on the subject, USTB was able to operationalize how many New Dealers felt, namely, that the federal government needed to be more and actively supportive of minorities. In this case, that meant encouraging African Americans to travel and have access to safe and welcoming tourist destinations.

Yet what USTB did was ever so slightly subversive to the racial status quo. Usually, New Deal programs that set aside a proportion of funding to assist African Americans focused on projects that largely maintained the social separation of whites and blacks. For example, as the WPA director for Indiana, Wayne Coy allocated some of its funding to build community centers in African American neighborhoods, improve school grounds in those locales, and hire women for sewing projects (Lee 2018, 44). These forms of aid were, of course, helpful and somewhat common in New Deal programs, but they were nonetheless premised on keeping the status quo of separation in place. On the other hand, USTB's efforts to promote more tourism by African Americans had the effect of trying to bring blacks into public spaces that were widely patronized by whites. For example, encouraging African Americans to visit national parks would lead to more public intermingling of whites and blacks. This meant that African Americans would be more visible publicly as an integral part of the American citizenry. Whites could not continue to wholly ignore their existence, if only by not being in their midst. While hardly revolutionary, USTB was taking a subtly subversive step toward increased social mingling in public venues. African Americans would no longer be invisible and segregated. That the bureau did this is to its historical credit.

Arguing over the Normative Role of the Federal Government in Travel

Arguments about the proper role of the federal government are as old as the Constitution itself. Early presidents often pondered about the precise and legal scope of the national government as explicitly stated, implied, or not mentioned one way or the other in the Constitution. What about federal funding to improve ports? Postal roads? To pay for the Louisiana Purchase? These arguments have continued nearly perpetually to the present day. Were gun-control laws constitutional? Could a president order the military to engage in combat without a declaration of war? Could a president target for assassination an American citizen abroad through a drone strike? Was requiring all states to participate in Obamacare an executive-power overreach?

The arguments in and out of Congress about the role of the federal government in tourism promotion present a synecdoche of this larger debate. Before, during, and after USTB's demise, Washington policy makers and media commentators had a running argument about the mission of the Travel Bureau (and its successors in Commerce). Some argued that the economic interests of tourism attractions did not rise to the level of a *national* matter. Sure, some state and local governments might have within their boundaries some locations that were highly dependent on tourism. If so, those governmental units might decide that it was in the public interest to spend tax dollars on promotion, such as in some more distant metropolitan areas where the population would not automatically think of visiting for vacations. Calling all New Yorkers: Have you considered a staying at a Montana dude ranch? Spending public funds for this was not only within their legal prerogative, it was also an application of their political and policy considerations. They were willing to spend tax funds based on what they interpreted as directly in their best (political and policy) interest. Sure, more jobs were a desirable general goal. For states with income taxes, that also meant a bump in tax revenues from those working in the sector. However, an even stronger justification was parochial self-interest. States, cities, and counties would benefit from increased revenues from tourists and travelers paying sales taxes, the gas tax, hotel taxes, and airport fees. To these politicians, such increased revenues were "free" in the sense that their constituents were not paying them. Out-of-staters were! Elected officials could argue, with some justification, that promoting tourism held down taxes that their voters would otherwise have to pay.

However, how was this a matter of responsibility for the federal government? For example, at a public hearing in 1931, Congressman Olger Burtness (R-ND) said, "The Federal Government as such, should not attempt to promote travel within different portions of the United States. . . . I do not think it is properly a governmental function, a Federal function, to encourage people to travel from one State to some other" (US House 1931, 40). Given that this comment was made before FDR's election, it was not a reflexive attack on the New Deal's expansion of the scope of the federal government. Rather, the observation was made by a Republican member of the majority party in Congress and when a Republican was in the White House. After FDR's election, in 1937, an editorial in a conservative Massachusetts newspaper opposed a federal travel agency because its premise would be that "the Government can do everything better than private groups" and because decisions about travel promotion should not reflect any particular national agenda and priorities set in Washington.[2] Two years later, another conservative newspaper editorialized that there was little need for such a federal agency.[3]

After WWII, Interior sought congressional funding to resume operations of the Travel Bureau. Congressman Robert Jones (R-OH), chair of the House Appropriations Subcommittee for Interior's budget said, "Do the railroads, busses, and other transportation travel groups not promote national parks and places of national historic interest in their own selfish desire to get people to travel? . . . I fail to see the reason why you [NPS/Interior] need a travel bureau" (US House 1947b, 617). A year later, Congressman John Heselton (R-MA) reacted to testimony at a public hearing by private travel businesses and professional associations in favor of federal spending on domestic travel promotion. He wondered aloud "just why is it that private interests involved here, who, it seems to me, do a very splendid job" were advocating that federal funds should be spent to do this (US House 1948e, 19).

These concerns were not limited to the question of initially establishing or funding USTB. Long after USTB had folded and its mission shifted to Commerce, various administrations continued raising this question. When President Carter vetoed a bill in 1980 to create an independent Travel and Tourism Administration, he said that the proper role of the federal government in this policy area should largely be limited to coordination of policy and information collection (Carter 1982, 2838). After President Reagan took office in 1981, Congress tried to pass modified legislation on the subject. Two committee members objected to the legislation in principle. Congressman Toby Moffett (D-CT) said the committee-recommended legislation made it appear as if there were "an immense, unmet need. Actually, nothing could be further from the truth," considering all the spending by the private sector and state governments (US House 1981, 29). James Collins (R-TX) similarly argued, "The travel industry is clearly capable of fulfilling its service [needs], without the Federal Government" (31). A year later, the Republican minority report on another related bill by the committee objected to the legislation because "we think that tourism promotion should more properly be a private sector program; the industry is well-organized and extremely successful so we see no real need for Federal intervention or support" (US House 1982, 7). In a separate dissenting statement, committee member William Dannemeyer (R-CA) said, "This is a question of delineating private sector versus public sector tasks, not a referendum on the importance of tourism to our economy" (8). As summarized in chapter 9, this yin and yang of the normative debate over a federal travel role have continued, whether a Democratic or Republican president, whether with a Democratic or Republican Congress.

That travel and tourism were not evenly spread throughout the country was oddly highlighted in a 1981 House committee report. In trying to make the case *for* greater federal tourism promotion, it stated that tourism was one of the three largest employers in thirty-eight states, implicitly acknowledging that it was not in twelve states. Another line of argument by the proponents was that tourism was the largest employer in fourteen states (US House 1981, 9). That meant it was not in thirty-six states. This economic data was a confession that travel and tourism were not national phenomena that were evenly spread out across the country. Therefore, taxpayers in both the thirty-six states where they were not the largest employers and the dozen states where they were not even in the top three employment sectors were exporting money to other states that would benefit from their largesse.

However, in reality, it was not much of a contest politically. Adapting Fehrenbacher's incisive observation regarding a very different and much more important federal policy debate (slavery), proponents of a federal tourism agency were an *interest*, while opponents were merely a *sentiment* (2001, 28). As with most policy areas, the political deployment of forces was asymmetrical. It was "the usual imbalance of passion and resources between special interests and the public" (Brill 2018, 205). The industry advocates cared deeply about only this one subject, while the proponents for a broad and diffuse public interest had to focus on everything. Therefore, opposition to actions that benefited a narrow special-interest group was merely a mile wide and an inch deep. Politicians could shrug it off. In American politics, the salience of a small group greatly outweighed politically the vague opposing interests of the public-at-large. Better to reward a small audience that is paying attention and cares a lot about this single issue than to fear punishment by the voters for doing so.

Once the tourism industry won the debate that the federal government should play an active role in domestic tourism promotion, the question moved on to money. For business, it was obvious that more was better. Sure, a $75,000 a year budget for USTB was good, but $100,000 would be better. More than that, best! In a sense, the industry's view was that there was no such thing as too much government spending on what it wanted. The key, of course, was that lobbying for higher funding had this focus because industry wanted to spend other people's money. For the travel business, using taxpayer dollars to pay for tourism promotion was free money. However, the industry was quite alarmed about being expected to pay for it. That would hurt profits and, besides, it was their own money.

What was the benefit of doing *that*? In 1990, a brief political compromise was that the funding would come neither from the for-profit travel business nor from the taxpayers at large. Instead, a one-dollar fee would be imposed on each passenger arriving in the United States on a foreign airline or ship.[4] It was the classic Washington maneuver of "Don't tax you, don't tax me, tax that fellow behind the tree."[5]

The key question was if the public-at-large obtained benefits and paybacks for spending tax funds on tourism promotion. It was a stretch of an argument. Sure, that increased tourism would benefit the for-profit companies getting more business was clear. However, a clear benefit for the national taxpayers was a stretch. Yet the kind of arguments made by the travel business was little different from what every special interest group said to Congress and presidents about increasing spending to benefit themselves. All organized interests claimed that the public would benefit if their little group prospered with the help of federal aid. The intense and self-serving interest by business was politically sanitized by the idea that USTB should have an advisory group from the industry. This could be justified as helping the bureau identify what was needed and how to go about getting it. Cooperation and communication were noncontroversial goals, long advocated by public administration—probably wrongly. The goal should be to identify and pursue the public interest, not special interests. Nonetheless, Ickes was delighted by the political support he was able to generate for USTB through its advisory board. It was a win-win situation.

However, inevitably, a special interest group wants to capture and coopt its agency. This inexorable arc of desired control came to a peak with the tourism promotion bill that President Carter vetoed in 1980. One of the reasons for his veto, he said, was that the federal government would be ceding control of a federal agency to the business sector, including control of congressionally appropriated funds. With the travel industry having a majority of members of the proposed oversight board, the new agency "would be more responsive to special industry interests than to the need for a coordinated Federal approach that will balance the needs of tourism against other national priorities" (Carter 1981, 2838).

∼

In all, the US Travel Bureau presents an interesting, if modest, story about the history of the federal government, FDR's presidency, and the practice of public relations as an element of public administration. Its record need not be exaggerated. Nonetheless, it was the first program and agency in the federal government with the goal of promoting domestic travel by the citizenry. It was a tangible example of the unprecedented expansion of the role of the executive branch during the New Deal; it made an unusual and pointed effort to promote equal travel opportunities to African Americans; and it was able to attract business support, creating a crack in the largely monolithic opposition by business to FDR and the New Deal.

USTB was also a different kind of federal agency. It provided no product or service other than engaging in public relations. The conservative coalition in Congress, a vehement and consistent opponent of agency PR, gave it a pass. Even for conservatives, USTB was doing good PR. After all, business liked it. This was no time to be doctrinaire. USTB's PR was also highly unusual in another sense. The informational products that federal agencies routinely disseminated were created in house. They were official products of the government and, as such, were assumed to be accurate, disinterested, and objective. Unemployment statistics, weather forecasts, and farming advice were always in the name of the government, specifically the voice of the neutral and merit-based civil service. Government was not taking sides, skewing research, or providing inaccurate data. And it never endorsed any particular business over another or locality over another. All were treated equally. This was a kind of public administration Good Housekeeping Seal of Approval. A citizen could always count on agency informational materials as impartial, high quality, and credible. Some of USTB's PR products fit this traditional template, but others did not.

As a way station for PR and marketing materials from self-interested groups, USTB was vaguely placing its imprimatur on such materials when it provided them to citizens. If a railroad brochure claimed it provided the best, fastest, and least expensive service to the West Coast, USTB was not interposing its judgement on that. In a sense, USTB was practically sanitizing such self-serving and for-profit information. If the federal government was distributing such a brochure, then the citizen might assume it carried the same credibility as brochures from other federal agencies. In truth, not. This odd and unusual PR role of USTB as a disseminator of self-interested advertising diluted the professional standing of the civil service system: No, we do not stand by the accuracy or impartiality of the material you receive from us. You cannot rely on something you get from us. For the core concept of a civil service dedicated to the public interest, this new kind of federal PR was *not* progress for the citizenry at large.

Long forgotten (or misremembered, as measured by the number of errors about it in the published literature), American history deserves to note the contributions of the US Travel Bureau and its unusual role in public administration, public relations, race relations, tourism, and business support.

Notes

Notes to Introduction

1. Associated Negro Press (ANP), "Gets Job with U.S. Bureau of Travel," *[Norfolk, VA] New Journal and Guide*, May 21, 1938, 3.

2. "An Act to encourage travel in the United States, and for other purposes," 54 *Stat.* 773.

3. These limitations and constraints delimiting authorization legislation versus appropriation legislation were no mere technicalities of parliamentary procedure. For example, FDR had to fight hard to get Congress to enact a law authorizing funding for the Office of Government Reports in the Executive Office of the President. Efforts to appropriate funding to it before such legislation had passed were ruled out of order (Lee 2005b, 67–79). More diffuse efforts by congressional conservatives to limit agency PR usually focused on amendments to appropriations bills because "provisos tucked into funding bills were easier to pass than a free-standing law" (Lee 2011, 173). However, any such limitations were for that fiscal year only and applied only to the agencies funded by that particular appropriations bill. (Generally, there were about a dozen separate appropriations bills that had to pass in order to fund the entire federal government for each fiscal year.) Many of these initiatives occurred after FDR, indicating a general congressional disposition rather than bias against Roosevelt. Lee (2011, chaps. 8–10) detailed prominent examples of post-FDR funding riders that set limits on agency PR spending.

4. Email from Jennifer Klang, Head of Reference Services, Library, Interior Department, January 19, 2018. Author's files. Also, the National Park Service's historian at the time was unable to suggest to her any other possible locations of those orders.

5. Emails from Dwight Pitcaithley, March 20, 2018, and Janet McDonnell, March 23, 2018. Author's files.

6. Email from Nancy Mulhern, Government Information Librarian, Wisconsin Historical Society, March 29, 2018. Author's files.

7. This phrase is usually attributed to Philip L. Graham, president and

publisher of the *Washington Post* (and married to Katherine Graham, who succeeded him). Fittingly (and apparently the first time he used it publicly) was when he said it in a speech to the 1953 annual conference of the American Society for Public Administration (ASPA) in 1953. "Public Administration and the Press," *Public Administration Review* 13:2, 88. The phrase apparently evolved from internal discussions by the *Post*'s editorial board, of which he was a member. In 1941–42, Graham had been a lawyer on the staff of Wayne Coy, head of the Office for Emergency Management, giving him personal experience with public administration (Lee 2018, 147). He then enlisted in the Army Air Corps (the precursor to the US Air Force) as a private, eventually rising to major.

Notes to Chapter 1

1. 71 H.R. 11431.
2. *Congressional Record* (*CR*) 72:6 (April 7, 1930) 6622–27.
3. 71 H.R. 13553. The Federation was an organization of US businesses interested in promoting travel. The "International" in its title is confusing, apparently intended to convey international travel rather than having international members or being an international organization.
4. *CR* 74:2 (January 5, 1931) 1397–98.
5. 72 H.R. 4676.
6. 72 H.R. 14178.
7. 74 S. 33 and 74 H.R. 5844.
8. Letter from Copeland to Senator W. G. McAdoo (D-CA), May 14, 1937. File: S. 3635, Box 79: S. 3635-S. 3706, Sen 75A–E1, Seventy-Fifth Congress, Committee on Commerce, Records of the US Senate (RG 46).
9. Another objection of the State Department to the bill obliquely referred to efforts by German Jews to escape Nazi Germany. At the time, the department made it extremely difficult for them to qualify for immigrant and other entry visas. By now, Hitler had been in power for 2½ years. The Department was concerned that a new law aimed at encouraging foreign tourism by reducing the bureaucratic obstacles to issuance of tourist visas would become a magnet for fraudulent visa applications because the traveler would have no intention of returning to Germany (US Senate 1935a, 4).
10. "Travel Boost," *Business Week*, August 3, 1935, 28.
11. "U.S. Travel Commission," *Traffic World* 56:1 (July 6, 1935) 6.
12. *CR* 79:11 (July 29, 1935) 11977.
13. *CR* 79:12 (August 16, 1935) 13352–53.
14. Barron C. Watson, "Federal Bid to Visitors," *New York Times* (*NYT*), February 9, 1936, XX-10.
15. *CR* 79:13 (August 23, 1935) 14324.

16. *CR* 80:1 (January 6, 1936) 112–13.
17. *CR* 80:5 (April 20, 1936) 5735; 80:6 (May 4, 1936) 6643.
18. "Expect Many to Travel in U.S.," *Altoona [PA] Mirror*, June 25, 1936, 17.
19. Letter from FDR to James Gerard, June 2, 1936. Box 133, General (0-201-13), Central Classified File 1933–49, National Park Service (RG 79), National Archives II, College Park, MD. All archival citations to follow are from this source, unless otherwise specified. Subsequent citations will list only the box number. The President's letter to Gerard was covered as news: "Park Publicity Move Hailed by Roosevelt," *NYT*, June 22, 1936, 19.

Notes to Chapter 2

1. Memo from Arthur E. Demaray, Acting Director, National Park Service (NPS), to Interior Secretary Harold Ickes, n.t., January 4, 1937. Box 133.
2. 49 *Stat.* 1894–95. Legislation is rarely immaculately conceived. Ickes had asked for it in 1934 (US House 1934, 2). USDA's Forest Service was explicitly excluded from the survey, even though national forests often encouraged recreational uses.
3. Loomis had previously been involved in the travel business in the private sector (McGarry 1936, 4). Earlier in his career, he had been in charge of recreational activities for soldiers at Fort Sill (OK). Delia Pynchon, "An American You Should Know," *Washington Star* (*WS*), December 29, 1938, A-9. In the mid-1930s, he was involved in creating Visitours, Inc., to "cater to the 'luxury' tourist trade" in New York City. Its office was located at the International Building of Rockefeller Center. "Business Firms in New Locations," *NYT*, May 15, 1935, 40. Because of his private-sector work, the New York State Chamber of Commerce appointed him to serve for several years on its tourist industry committee. "Chamber Picks Board to Study Corporations," *NYHT*, June 23, 1939, 27. For a photo of Loomis, see Scott Hart, "Federal Diary," *WP*, August 11, 1938, 9.
4. "Memorandum for the Press; For Release February 11, 1937," n.t. Box 134. The title of the new unit was in lowercase in the press release, implying that the title was more descriptive of its function than its formal and official name.
5. A. Sportsman [pseudonym], *Outdoor Activities* (column), *Los Angeles Times*, February 21, 1937, F2.
6. "Out-o-doors" (column), *Charleston [WV] Gazette*, March 28, 1937, 15.
7. "Federal Tourist Bureau Is Set Up Here as Ickes Promotes 'See America' Movement," *NYT*, March 30, 1937, 25.
8. United Press, "Vacation Advice Given by Uncle Sam," *Louisville [KY] Courier-Journal*, June 27, 1937, 29; International News Service, "Federal Tourist Bureau Will Stimulate Industry," *San Bernardino [CA] County Sun*, July 6, 1937, 9; "Federal Tourist Bureau," *NYT*, May 16, 1937, XX-3.
9. Press release (n.t.), April 4, 1937. Box 134.

10. For a photo of the Broadway office, see Margaret F. Ryan, "In Other Government Agencies: National Park Service," *School Life* 23:3 (November 1937) 93. *School Life* was published by the US Office of Education, then a bureau in the Interior Department.

11. "Uncle Sam a Tourist Agent" (editorial), *Nashville [TN] Banner*, October 1, 1937. Box OV 26, Ickes Papers. The efforts of the USTB might encourage those "dashing madly to Florida" to become aware of "various points of interest en route," such as in, ahem, Tennessee.

12. To post–WWII sensibilities, the title sounds odd, even seditious. It must be remembered that the widespread negative meaning of the word occurred after WWII, particularly to characterize French citizens who cooperated with the Nazi and Vichy regimes. When used by USTB, the title and role of collaborator did not have such negative connotations.

13. Memo from NPS Director to Ickes, July 24, 1937. Box 133.

14. Sunday, September 12, 1937: "J. W. Gerard to Hold Federal Park Post," *NYT*, N1; "Gerard to Help Tourist Bureau Promote U.S. Travel Virtues," *Washington Post* (*WP*), 4. There were also two articles in that day's *Washington Star* (*WS*): a news story, "Gerard Named," A-9, and a column in the travel section, Jacques Futrelle, Jr., "Traveler's Notebook," B-5. Later coverage included Associated Press (AP), "See America First," *Christian Science Monitor* (*CSM*), September 13, 1937, 4; "The United States Tourist Bureau," *American Foreign Service Journal* 14:10 (October 1937) 593, 620.

15. Text of radio address, October 13, 1937. Box 134.

16. Richard W. Dunlap, "Part of the Task of Re-Evaluating America Is Visiting Its Many Diverse Areas," *New York Herald Tribune* (*NYHT*), October 10, 1937, J31.

17. "Gerard and Foreign Tourists" (editorial), *Worcester [MA] Telegram*, September 18, 1937. Box OV 25, Ickes Papers.

18. "A Bureau Boost" (editorial), *Greenville [TX] Banner*, November 27, 1937. Box OV 26, Ickes Papers.

19. "Familiarity Breeds Good Will" (editorial), *Sterling [IL] Gazette*, December 20, 1937. Box OV 27, Ickes Papers.

20. "Uncle Sam, Travel Guide" (editorial), *Springfield [MA] Union*, December 14, 1937. Box OV 27, Ickes Papers.

21. Marshall Sprague, "Uncle Sam, A Travel Aid," *NYT*, September 19, 1937, XX-2.

22. US Department of Interior, Press Memorandum, November 16, 1937. Box 134.

23. "The United States Tourist Bureau" (typescript), n.a., n.d. (probably November 1937). Box 134. This internal document was in part a quarterly summary of activities as well as plans for the near future. It probably also served as an early draft for the brochure about the new bureau published by the Department in December (USTB 1937b).

24. Memo from Loomis to Isabelle F. Story, NPS, August 18, 1937. Box 134.

25. Intense and ongoing congressional efforts to hold down agency printing budgets was a longstanding generic focus on Capitol Hill, originally triggered by the high cost of publications of the US Geological Survey, a different bureau in Interior (Lee 2011, 30–31, 137–39).

26. Letter from American Express Company to Arno B. Cammerer, Director, NPS, July 13, 1937. Box 133. American Express viewed the tourist promotion work of other countries as "scientifically maintained" and that the United States needed to do the same. "New U.S. Tourist Bureau will Help Travel Industry," *Hartford [CT] Courant* (*HC*), June 20, 1937, B6.

27. "Program Mapped to Draw Tourists," *NYT*, December 12, 1937, N-1, N-3.

28. Greeley had been the editor of the *New York Tribune* in the mid-1800's. The paper's annual civic forum was intended as a serious venue for discussion of public policy, as well as a prestige showcase for the paper and its self-appointed role of respectable Republicanism. That FDR made such a frivolous comment was indirectly a political poke in the eye. He would not provide the paper his imprimatur of seriousness and importance. Yet, by agreeing to provide a greeting, he finessed a potential criticism that he was boycotting the event as small-minded political payback for the paper's opposition to him.

Notes to Chapter 3

1. Memorandum for the Washington Office, from NPS Director Cammerer, March 3, 1938, document file no. 17317. Box 134.

2. Memo from Horton S. Allen, Acting Supervisor, to Mr. [Conrad] Wirth, "Weekly Report—Week Ending August 26, 1938." Box 133.

3. "U.S. Tourist Bureau" (organization chart), n.d. Box 133.

4. "Report of Activities," *Official Bulletin of the United States Travel Bureau—National Park Service* (hereafter *OB*) 1 (October 1938) 2.

5. "U.S. Gives Nature a Hand in Restoring Perspective," *WP*, October 30, 1938, TS-10.

6. Press release from NY Field Office, n.t., n.d. (about June 8, 1938), Box 134. The change attracted little press attention: "Agency Changes Name," *Chicago Defender* (national weekly edition), June 18, 1938, 7; "Change in Name," *Detroit Free Press*, June 19, 1938, 4–8. FDR seemed to like the construction "United States ____ ____" for agency names, such as United States Film Service (Lee 2005, 27–29), United States Information Service (17), and United States Information Center (118). The prefix "United States" helped convey to citizens that an agency and its programs came from the federal government.

7. "Government Tourist Bureau Will Aid the Travel Industry," *Boston Transcript*, June 19, 1938. Copy located in internal memo, document file no. 27019. Box 134.

8. "Tourist Tide" (editorial), *Boston Globe*, July 29, 1940, 10. Perhaps adding to the confusion, as late as 1940, the bureau kept using printed stationery with its old name in the letterhead. The word "Tourist" was stamped out and "Travel" stamped above it. Letter from Horton S. Allen, Jr., Acting Supervisor, New York Office, USTB to Public Documents Library, GPO, Washington, DC, February 7, 1940. File: Sudoc I 29.37 (USTB), Box 1, Interior Department, Publications of the Federal Government (RG 287).

9. Letter from Gerard to FDR ("My dear Franklin"), June 13, 1938. File: Gerard, James W., Box 136, Subject File, PSF 1933–1945, FDR Library. The archival collection of the library did not have any reply from FDR to Gerard upon his resignation. Email from Virginia Lewick, FDR Library Archivist, to the author, August 14, 2018. Author's files. Given that it was Ickes who had named Gerard as collaborator, perhaps FDR felt that any thank you for his USTB service should more logically come from Ickes.

10. "Tourists Favor National Parks, Historical Shrines and Resorts," *NYHT*, January 29, 1939, D11.

11. Ibid.

12. His legal name was James Lee Bossemeyer, but professionally he usually identified himself as J. Lee Bossemeyer or J. L. Bossemeyer.

13. "Travel Bureau Head Discusses National Parks," *Berkeley [CA] Daily Gazette*, November 5, 1938, 8.

14. Jack Burroughs, "Travel Confab," *Oakland [CA] Tribune*, October 7, 1938, 37.

15. N.t., *[Reno] Nevada State Journal*, March 13, 1938, 4.

16. Robert L. Spencer, United Press, "Winter Sports Sweep Country . . . Millions Play in National Parks," *Berkeley [CA] Daily Gazette*, January 19, 1938, 10.

17. "Civic Groups Join Forces to Boost West," *Salt Lake [City, UT] Tribune*, January 11, 1938, 1.

18. "Montanans Officials to Aid Dude Ranchers," *[Butte] Montana Standard*, December 9, 1938, 11.

19. "Travel Bureau Head Discusses National Parks," *Berkeley [CA] Daily Gazette*, November 5, 1938, 8.

20. "Ships at Sea Today Carrying Hundreds on Holiday Cruises," *NYHT*, December 25, 1938, D11; "U.S. Park Officials Will Speak Here," *Atlanta Constitution*, February 6, 1938, 2A.

21. "More Travel Here" (editorial), *Boston Globe*, April 9, 1938, 14.

22. *WP*: Scott Hart, "Federal Diary" (daily column), June 5, 1938, TT8; "Travel Notes," June 12, 1938, TT11; Arthur Ellis, "Camera Angels," August 7, 1938, TT10.

23. Interior Department press release (n.t.), August 5, 1938. Box 134.

24. Memo to Mr. [Conrad] Wirth from Acting Chief, Research Unit, n.t., August 9, 1938, document file no. 1922019. Box 133.

25. North American Newspaper Alliance (news service syndicate), "American Travel Boom in Prospect," *WS*, April 17, 1938, C-11; Christopher Janus, "The Tourist Is Analyzed," *NYT*, August 21, 1938, XX-1; NPS press release, n.t., November 25, 1938, Box 134.

26. A cover letter from Wirth indicated that the report was being given limited circulation as a prototype and was likely incomplete. He invited readers to submit additional relevant information to be incorporated into a more formal report for broad release the next year. Wirth cover letter, October 18, 1938. Box 134.

27. Address by Nelson Loomis, September 26, 1938. File: Sudoc I 29.37 (USTB), Box 1, Interior Department, Publications of the Federal Government (RG 287).

28. Photos, Travel Bureau Conference, December 17, 1938. Accession nos. 2013-1116 and 2013-1117, Truman Presidential Library, accessed January 7, 2020, https://www.trumanlibrary.org/photographs/view.php?id=42133 and https://www.trumanlibrary.org/photographs/view.php?id=42157.

29. Interior's press office released a predelivery transcript of his radio address to assist reporters covering it. Memorandum, n.a., n.d., n.t. Box 133.

30. Letter from FDR to Ickes, January 14, 1938. Box 134.

31. "Ickes on Radio to Spur Nation's Tourist Bureau," *NYT*, January 16, 1938, 12; AP, "Ickes Invites Tourists," *CSM*, January 18, 1938, 13; "Roosevelt [sic] Speech Launches Radio Drive for Tourists," *WP*, January 19, 1938, X4.

32. Entry for January 18, 1938, Ickes Diaries (unpublished version), p. 2568.

33. "Tourist Travel Topic," *WS*, April 1, 1938, A-3. A week later, an article in the *Times* summarized an extensive "statement" by Gerard about USTB's plans. It is possible that the statement was a print version of his radio script, perhaps rereleased as a formal press release. "U.S. Plan to Spur Travel Described," *NYT*, April 8, 1938, 21. The article was not identified as coming from the paper's Washington bureau, suggesting that it had been issued by the USTB's New York office and therefore treated by the *Times* as local news.

34. "International Networks," January 31–May 16, 1938. Box 134.

35. "Radio Travelogues Scheduled Weekly," *Regional Review* (NPS Region 3 newsletter) 1:2 (August 1938) 3; "Today's Radio Programs," *NYHT*, August 26, 1938, 32; "Today on the Radio," *NYT*: September 9, 1938, 24; September 23, 1938, 41; September 30, 1938, 38; October 7, 1938, 20.

36. Transcripts: "Your Government at Your Service," August 12, 1938; "Question Box Program," November 9, 1938. Box 134.

37. "Tourism" (editorial), *New Orleans [LA] Item*, January 21, 1938. Box OV 29, Ickes Papers.

38. "Seeking Travel Balance" (editorial), *Miami [FL] Herald*, January 23, 1938. Ibid.

39. "U.S. Travel Bureau" (editorial), *Los Gatos [CA] Mail-News*, January 27, 1938. Ibid. "Spurring Travel in America" (editorial), *Fort Wayne [IN] Journal Gazette*, April 10, 1938. Box OV 31, Ickes Papers.

40. PWA was a freestanding federal agency (often confused with WPA) and was funded by a separate appropriation. Ickes headed it, but wholly apart from his "hat" as Interior secretary. The difference between WPA and PWA was that WPA employed people directly (whether for infrastructure projects or art), while PWA gave grants to state and local governments to fund infrastructure construction, thus creating jobs but not employing the workers directly.

Wearing more than one hat, as Ickes did at PWA and Interior, was relatively common in FDR's presidency. For example, William McReynolds was the head of the Liaison Office for Personnel Management, administrative assistant to the president, and the liaison officer for the Office for Emergency Management. He was succeeded in the emergency management position by Wayne Coy, who simultaneously was special assistant to the president. Later, Coy also wore the hat of assistant director of BOB. Coy's emergency management successor was (briefly) James Byrnes, who also wore the hat of director of the Office of War Mobilization (Lee 2016; 2018). During WWII, Ickes wore other hats, such as petroleum coordinator.

41. Transcript of "Radio Forum Discussion: Economic and Cultural Aspects, Domestic and Inter-American Travel," November 14, 1938, document file no. 42085. Box 133.

42. Entry for November 15, 1938, Ickes Diaries (unpublished version), p. 3042.

43. *WS*: "Forum on Inter-American Travel to Dedicate Interior Department Broadcasting Studio," November 13, 1938, A-2; "Forum Speakers Tell of Plans to Push American Travel," November 15, 1938, A-15.

44. "Ickes Speaks for Inter-American Travel," *Travel Trade* 19:5 (December 1938) 25.

45. Cover letter, "Important Notice," n.d. (probably late November 1938), document file no. 42381. Box 133.

46. Robert L. Perry, travel editor, "National Parks Beat '37 Pace," *Detroit Free Press*, August 21, 1938, 1–10. The paper ran in its Sunday travel section several USTB releases that promoted national parks: "All Western National Parks Open for Mid-June Travelers," June 5, 1938, 1–16; "Zion Canyon Vista Dazzles by Colors," July 3, 1938, 1–13; "Waterfalls to Dwarf Niagara Give Yosemite a Place Apart," July 10, 1938, 1–14.

47. Nelson A. Loomis, "Quaint U.S. Areas Await Discovery," *WS*, December 11, 1938, G-4.

48. *NYHT*: "U.S. Tourist Office Gets Aid of States and Private Units," April 24, 1938, D9; Dunlap, "Wherever the Vacationist Goes, a National Park Is Not Far Away," September 25, 1938, D10; "Ships at Sea Today Carrying Hundreds on Holiday Cruises," December 25, 1938, D11. *WS*: Futrelle, "Traveler's Notebook," September 11, 1938, C-7, November 20, 1938, B-5; Russell E. Singer, "1938 Motor Travel Is Near Record Set during Last Year," November 6, 1938, G-14. Sunday (or weekend) coverage in other papers: Elizabeth C. Taylor, "Range of Tourists Has Grown Vastly," *WP*, July 31, 1938, B5; Christopher Janus, "The Tourist Is Analyzed,"

NYT, August 21, 1938, XX-1; "Tourist Bureau Attracts Foreign Visitors to U.S.," *CSM*, May 14, 1938, V5.

49. Arthur Ellis, "Camera Angles" (column), *WP*, August 7, 1938, TT10.

50. The name of the magazine reflected that its leadership came from two prominent radio commentators: Lowell Thomas and H. V. Kaltenborn.

51. "Information for Tourists," *Journal of Adult Education* 10:2 (April 1938) 214.

52. "Let Us Hear About It!!" *OB* 1 (October 1938) 4.

53. N.t., *OB* 2 (November 1938) 4.

54. Associated Negro Press (ANP), "Gets Position with U.S. Travel Bureau," *Detroit Tribune*, May 14, 1938, 1; "Get Diplomas from Ethical Culture School," *New York Amsterdam News*, June 11, 1938, 10.

55. George S. Schluyler, *Views ReViews* (weekly column), *Pittsburg Courier*, July 30, 1938, 10.

56. "McDowell, 69, Travel Agent, Dies" (obit.), *Chicago Defender* (national weekly edition), December 26, 1953, 4.

57. "Race Travel is Helped by Tourist Aid," *Chicago Defender* (national weekly edition), June 18, 1938, 17 (women's page).

58. "Naming Tourist Bureau Collaborator, Ickes Seeks to Aid Race Travelers By [sic]," *Chicago Defender* (national weekly edition), May 28, 1938, 2.

59. ANP, "Gets Job with U.S. Bureau of Travel," *[Norfolk, VA] New Journal and Guide*, May 21, 1938, 3.

60. Schluyler, op. cit.

61. "Federal Board to Aid Tourists in U.S. Planned," *NYHT*, September 28, 1937, 31.

62. Memo for the Press, n.t., December 14, 1937. Box 134. The curator of the Department of Greek and Roman Art at the Metropolitan Museum of Art in New York responded to Gerard's mailing by writing directly to Senator Copeland to express support for his legislation. Letter from Giseta M. A. Richter to Copeland, April 11, 1938. File: S. 3635, Box 79: S. 3635-S. 3706, Sen 75A–E1, 75th Congress, Committee on Commerce, RG 46.

63. 75 S. 3018. AP, "Bill Asks Tourist Board," *NYHT*, November 17, 1937, 2.

64. "Tourist Trade for the U.S." (editorial), *Southbridge [MA] News*, December 13, 1937. Box OV 27, Ickes Papers.

65. 75 S. 3635.

66. Astonishingly, the State Department reiterated that one of its objections to the bill continued to be the issue it had raised in 1935 (chap. 1). Making it easier for Europeans to obtain tourist visas to the United States could have the effect of opening the doors to travelers who had no intention of returning to their country of origin but who could not obtain an immigration permit (read: German Jews). Between the department's January 1935 and April 1938 letters stating this concern about the bill, the Nazi regime had adopted the Nuremberg laws and many other discriminatory policies excluding German Jews from almost all aspects

of life, employment, education, and citizenship. Nonetheless, the State Department's objections still stood unchanged.

67. Letter from E. K. Burlew, Acting Secretary of the Interior, to Copeland, May 18, 1938. File: S. 3635, Box 79: S. 3635-S. 3706, Sen 75A-E1, 75th Congress, Committee on Commerce, RG 46.

68. *CR* 83:8 (June 9, 1938) 8633; (June 13, 1938) 8991–92.

69. The reason for the committee's inaction is undocumented. The committee's bill file at the National Archives' Center for Legislative Archives (CLA) contained no helpful information or correspondence.

Notes to Chapter 4

1. "Travel Conference Held in Washington," *OB* 4 (January 1939) 1; "Conference Indorses [sic] Travel Bill," *OB* 5 (February 1939) 1, 3.

2. "Increase of Travel to South America, West Indies Shown," *NYHT*, November 12, 1939, D12.

3. "U.S. Travel Bureau Here Busy," *NYHT*, May 14, 1939, D12; "Federal Diary" (column), *WP*, July 6, 1939, 24.

4. Futrelle, "Traveler's Notebook" (column), *WS*, September 10, 1939, C-9.

5. "Exposition Closes October 29th," *Travel News* [publication of USTB's San Francisco office], October 1, 1939, 1. This was not exclusively due to the war; financing the cost of a second year of operations was also a problem.

6. Futrelle, "Traveler's Notebook" (column), *WS*, February 4, 1940, D-11; "Western Travel Bringing Visitors to Boulder Dam," *CT*, February 18, 1940, G6.

7. "America's Attractions Unsurpassed, First Lady Says," *Travel and Recreation News Letter* [publication of USTB's NY office], 4 (October 3, 1939) 1.

8. "New Miami Boom Traced to European War," *Travel and Recreation News Letter* 5 (October 20, 1939) 1.

9. Ickes, Text of Address to AAA, November 16, 1939. File: Speeches #238: Am. Automobile Assn., Box 318, Ickes Papers. Macnamee saw an early draft of the speech and asked Ickes to remove unrelated swipes at the American Hotel Association so as not to undermine its support for the USTB legislation. Letter from Macnamee to Ickes, November 13, 1939, File: Drafts, Speeches #238, Box 320, Ickes Papers. Indeed, the delivered version of the speech omitted any critical references to the hotel organization.

10. "Road Hogs Hit by Ickes in A.A.A. Talk," *WS*, November 16, 1939, B-1.

11. "A.A.A. Rejects Ickes Plea to Limit Trucks," *WP*, November 18, 1939, 3; "Truck Men Dub Ickes Lord of Mistaken Ideas," *CT*, November 19, 1939, B9.

12. Ickes Diaries, unpublished version, November 19, 1939, p. 3910.

13. "Travel Booster," *Detroit Free Press*, March 26, 1939, 4.

14. Mrs. Roosevelt later claimed that her suggestion to Ickes was made in response to a request from FDR that she do so and that passing along Rohde's inquiry to Ickes was not a recommendation per se. Confusingly, Ickes replied that "the President *spoke* to me about Mrs. Rohde" (emphasis added), rather than Mrs. Roosevelt writing him. Letter from Mrs. Roosevelt to Ickes, April 4, 1939; reply by Ickes to Mrs. Roosevelt, April 8, 1939. File: Appointments 1938–1942, Box 97, Ickes Papers.

15. AP, "Mrs. Rohde Named to $1-A-Year Post," *NYT*, March 26, 1939, G-6.

16. "Mrs. Rohde Joins Travel Bureau Staff," *OB* 7 (April 1939) 1.

17. "First Inter-American Travel Congress," *Press Releases [of the State Department]* 20:496 (April 1, 1939) 258. A few months later, this official publication was renamed *The Department of State Bulletin* and issued monthly.

18. "Ruth Bryan Rohde Due Here Wednesday," *Los Angeles Times* (*LAT*), April 13, 1939, A5; "Ruth Bryan Rohde Speaks to Travelers' Aid Friday," *CT*, April 24, 1939, 13.

19. Rohde, "US Travel Bureau Filling Long-Felt Need for Service," *CSM*, June 8, 1939, F8.

20. In the 1920s, Macnamee had been a reporter for Universal Service, the Hearst newspaper empire's news service for its morning newspapers. In 1933, he became the director of the public relations division of FDR's National Recovery Administration (NRA). Pendleton Herring, one of the first political scientists interested in FDR's expansion of public relations in public administration, invited Macnamee to be one of four practitioners at a session on "Public Relations of National Administrative Agencies" at the 1934 annual conference of the American Political Science Association. It was a who's who of government PR practitioners and observers. The latter category included professors Marshall Dimock and John Gaus, Louis Brownlow of the Public Administration Clearing House, and private-sector practitioner Edward Bernays (Ogg 1935, 109). A few years later, Macnamee became a special assistant to Ickes in Ickes's (separate) capacity as head of PWA. *Congressional Directory*, 76th Cong., 1st sess., 1st ed., corrected through December 20, 1938, p. 385. While his name was sometimes capitalized in publications as MacNamee, this was incorrect.

21. Ickes, unpublished diaries, March 25, 1939, p. 3327.

22. In 1951, Congress legally prohibited federal agencies from trying to lobby Congress indirectly by urging private groups and citizens to contact their legislators (Lee 2011, chap. 9).

23. After his departure from the UTSB, Loomis joined the staff of the Republican National Committee as its Western field representative. "Reno Republicans May Hold Dinner," *[Reno] Nevada State Journal*, January 30, 1940, 3; "Dinner Opens G.O.P. Drive," *Oakland [CA] Tribune*, February 12, 1940, 4-D; "G.O.P. Hope Told Leaders," *LAT*, February 28, 1940, II-1. Four years later, he was the assistant

director of the Dewey presidential campaign for Southern California. "Dewey Office Ready to Open," *LAT*, August 27, 1944, 4. Loomis died in California in 1948.

24. "First Lady Intervenes," *Detroit Free Press*, June 18, 1935, Part 1, 5.

25. *WP*, June 18, 1939, 11.

26. *[Lincoln, NE] Sunday Journal and Star*, June 18, 1939, 1-A, 4-A.

27. *Detroit News*, June 18, 1939. Box 134.

28. *San Antonio [TX] Express*, June 18, 1939, 1, 5. At the time, newspaper headlines, always seeking brevity, sometimes identified President Roosevelt as "F.D."

29. *Kansas City Kansan*, June 18, 1939. Box OV 41, Ickes Papers.

30. June 23, 1939, 3.

31. In July, Mrs. Rohde was still mentioned in news coverage in her role with USTB: Pauline Frederick (NANA), "Mrs. Rohde Thinks a Woman Will Be President Someday," *WS*, July 23, 1939, C-8; "Mrs. Ruth Rohde Is Writing Book," *WP*, July 25, 1939, 10. Even more oddly, in December, more than half a year after she had departed USTB, the newsletter of its New York office published a column she had written: Rohde, "Why a Federal Travel Bureau?" *Travel and Recreation News Letter* 8 (December 5, 1939) 3. Equally surprising and inaccurate, in January 1940, the *Star*'s reader Q&A carried this exchange: "Q: What is Ruth Bryan Rohde's position in the Government? A: Mrs. Rohde is collaborator of the United States Travel Bureau." Frederic J. Haskin, "Haskin's Answers to Readers' Questions," *WS*, January 5, 1940, A-8. Apparently, she continued listing her USTB appointment in her *résumé* without stating that it had ended. In 1941 and even in 1947, newspapers still referred to her in the present tense as USTB collaborator. "Peace Needs Police to Enforce Law, Former Woman Envoy to Norway [sic] Says," *CSM*, May 22, 1941, 5; "Ruth Bryan Owen to Lecture Monday at MSC [Mississippi Southern College] Chapel," *Hattiesburg [MS] American*, January 31, 1947, 11. A biography of Rohde did not mention her USTB role. Sarah Pauline Vickers, *The Life of Ruth Bryan Owen: Florida's First Congresswoman and America's First Woman Diplomat* (Tallahassee, FL: Sentry Press, 2009).

32. Harlan Miller, "Over the Coffee" (column), *WP*, July 14, 1939, 2; "FBI Hunt for Spies Centered in Hawaii," *Oakland [CA] Tribune*, July 23, 1939, B-9.

33. "Federal Diary" (daily column), *WP*, August 25, 1939, 30.

34. Memo from Macnamee to Miss Story, December 2, 1939. Box 134.

35. "Notice of Presidential Project Authorization, Fiscal Year 1940," July 7 and September 20, 1939. Box 134.

36. For example, "A Guide to Alaska," *OB* 20 (December 25, 1939) 12.

37. "A Presidential Proclamation," *OB* 3 (December 1938) 1.

38. Common usage at the time treated the term "newsletter" as two separate words. Inconsistently, issues of the *News Letter* were numbered, but those of *Travel News* were not.

39. "Announcement," *OB* 11 (July 25, 1939) 3.

40. New York Office: "Travel through Tennessee to Southwest an Interesting Route," November 14, 1939; San Francisco Office: n.t. (on the upcoming Inter-American Travel Congress), March 23, 1939. Box 134.

41. "The United States Travel Bureau Serves the Camper and Tourist," *Camping Magazine* 11:1 (January 1939) 22–23.

42. "More N.E. [New England] Travel Bureau's Forecast," *Boston Globe*, June 4, 1939, B18; Dunlap, "New Map Shows All Park Sites and Describes Their Attractions," *NYHT*, September 24, 1939, D11; "Tenderfoot Tales," *Atlanta Constitution*, October 15, 1939, 6B; Oliver McKee Jr., "'Good Will' Spread," *NYT*, October 22, 1939, XX-1.

43. "Vacationists Spend Little for Food and Quarters," *WP*, July 20, 1939, 9; "Rates Michigan Seventh," *Detroit Free Press*, July 30, 1939, Part 1, 17; "Tourists Spend Billions Yearly," *Detroit Free Press*, September 17, 1939, Sports Section, 8; "Bureau Completes State Travel Promotion Survey," *OB* 20 (December 25, 1939) 4.

44. "New Trails Are Available for 'Pack' Hikers," *Berkeley [CA] Daily Gazette*, December 13, 1939, 19.

45. Bruce MacNamee [sic], "U.S. to Widen 'See America' Travel Drive," *WP*, October 15, 1939, A9.

46. US Office of Education, "School Tours," Circular No. 177, January 1939, document file no. 48986.

47. "Increase in Requests Received by the Bureau's New York Office," *Travel and Recreation News Letter* 10 (January 20, 1940) 7.

48. "N.Y. Fair Exploits the National Parks," *Detroit Free Press*, May 14, 1939, Part 1, 21.

49. "Working in Interest of Race," *Chicago Defender* (national weekly edition), April 29, 1939, 6.

50. McDowell, "U.S. Aids Travel among Negroes," *Evansville [IN] Argus*, September 30, 1939, 6.

51. "Increase in Requests Received by the Bureau's New York Office," *Travel and Recreation News Letter* 10 (January 20, 1940) 7.

52. "Cardozo Club to See Fair," *Baltimore Afro-American*, May 20, 1939, 6.

53. "Leaders Hail Business Planning Confab," *Chicago Defender* (national weekly edition), March 4, 1939, 6.

54. ANP, "Dixie Fair Tour Sponsored in S.C.," *Atlanta Daily World*, August 28, 1939, 2.

55. ANP, "Marie Dressler Funds Open Ga. Tourist Home," *Philadelphia Tribune*, November 16, 1939, 1.

56. McDowell, "Travel Promotion among Negroes," *Travel and Recreation News Letter* 1 (August 19, 1939) 7.

57. McDowell, "Travel Notes," *Travel and Recreation News Letter* 5 (October 20, 1939) 4.

58. Ibid. 6 (November 6, 1939) 8.

59. Ibid. 7 (November 20, 1939) 8; 8 (December 5, 1939) 8. An odd item in a September issue (not explicitly authored by McDowell) was of the Mississippi Advertising Council issuing a pamphlet (apparently) to encourage African American tourism, especially to Mound Bayou, "the first all-Negro town in the South." It had been founded by Isaiah Montgomery, a former slave of Confederate president Jefferson Davis. "Mississippi in Tribute to its Negro Population," *Travel and Recreation News Letter* 3 (September 19, 1939) 6. The only known copy of the pamphlet, "Songs of the Soul," is in the Special Collections of the University of Mississippi's J. D. Williams Library (OCLC #66391422).

60. 76 S. 307 and 76 H.R. 1792.

61. "Congress Bills Ask New National Travel Agency," *Automobile Topics* 133:3 (February 13, 1939) 85.

62. Ickes unpublished diary, January 29, 1939, p. 3195. For a photo, see "Conference Indorses [sic] Travel Bill," *OB* 5 (February 1939) 1.

63. "Solidarity Program Boosted by Ickes' Travel Proposal," *WS*, January 25, 1939, A-2; "Ickes's Promotion of Touring Backed," *NYT*, January 26, 1939, 10; "Business Indorses [sic] U.S. Travel Bureau," *WP*, January 26, 1939, 5; M. Le Tour (pen name), "U.S. Travel Bureau in Prospect as Bill Is Pressed in Congress," *CSM*, January 31, 1939, 5.

64. "Travel Legislation Indorsed [sic] by Press," *OB* 6 (March 1939) 2.

65. L. C. Speers, "Bill to Aid Tourist," *NYT*, February 5, 1939, XX-1.

66. Scott Hart, "Federal Diary," *WP*, February 22, 1939, 24.

67. "Bailey Advertises the State" (editorial), *Ashville [NC] Times*, January 6, 1939. Box OV 38, Ickes Papers.

68. "Another Bureau" (editorial), June 3, 1939. Box OV 40, Ickes Papers.

69. Ray Tucker, "National Whirligig" (syndicated column), *Iowa City Press-Citizen*, January 7, 1939, 4.

70. Even though it "had no objection" to the bill, the State Department continued to express its concerns that any new policies to make it easier for Europeans to obtain tourist visas "might interfere with our immigration practices," i.e., preventing German Jews from entering the United States on tourist visas, but with no intention of leaving. Letter from Secretary of State Cordell Hull to Senator Josiah Bailey, Chair, Committee on Commerce, February 14, 1939. File: S. 309, 76th Congress, Records of the Committee on Commerce and Related Committees 1816–1968, US Senate Records (RG 46). This position was a continuation of the Department's stated policy in 1935 and 1938 (chaps. 1 and 3). In this case, the restatement of its position (and the continuing relevance of the policy) came after the appalling Kristallnacht pogrom in November 1938 and Hitler's speech to the Reichstag on January 30, 1939, that should war come, it would lead to "the annihilation of the Jewish race in Europe."

71. 76 H. R. 5412.

72. The subject was considered so uncontroversial and the hearing so minor that the transcript of the hearing was not published, as is routine for most congressional hearings.

73. The formal letter from Commerce to the committee concluded with the careful statement that "the Bureau of the Budget has advised [the department] that there is no objection of the transmittal of this report to your Committee" (8). In other words, it was okay by BOB for Commerce to state its opposition to the bill publicly but that this was *not* an endorsement of the department's position by BOB's legislative clearance office.

74. Letters from Senator Alvin Victor ("Vic") Donahey (D-OH) to Hopkins and Loomis, June 7, 1939. File: S. 307, op. cit.

75. A related common phenomenon that politicians often experience is that losers tend to remember longer than winners, bearing a grudge while the winners move on.

76. Letter from Hopkins to Senator Donahey, June 27, 1939, File: S. 307, op. cit. A carbon copy of that letter was in USTB's office files. This probably meant that Hopkins directed his secretary to send Ickes a copy to document for Ickes that Hopkins had signed and sent it (Box 134). Oddly, an additional round of correspondence between the Senate and the Commerce Department indicated just how tenuous the department's endorsement of the bill was. Only six weeks after Hopkins's letter, his administrative assistant replied to a follow-up Senate inquiry that "the matter is now under consideration and you will be advised as soon as possible." Letter from M. Kerlin, Administrative Assistant to the Secretary, to Senator Burton Wheeler, Chair, Committee on Interstate Commerce, August 5, 1939. File: H.R. 6884, op. cit. Under routine circumstances, once an agency's position on legislation had been determined and the precedent set, later correspondence would simply and promptly restate it. It is possible that Hopkins was being pressured from within the department to rescind his earlier position. At the very least, it indicates that Hopkins's absences led to substantial lags between important incoming mail and his administrative assistant's ability to reply promptly on Hopkins's behalf.

77. Hopkins's frail health finally led to him resigning as secretary in September 1940. Now living full-time in the White House, he became FDR's emissary first to Churchill and then to Stalin in FDR's pre–Pearl Harbor efforts to oppose Hitler. Beginning in March 1941, Hopkins formally headed Lend-Lease, though again displaying his lack of interest in management. Giving up that official duty in August 1941, he continued as the unofficial policy maker for Lend-Lease through the war. He died in 1946.

78. 76 H. R. 6884.

79. *CR* 84:10 (July 31, 1939) 10551–53.

80. Bruce Macnamee, "U.S. to Widen 'See America' Travel Drive" (column), distributed by United Press, *WP*, October 15, 1939, A9.

81. The first session of the Seventy-Sixth Congress adjourned on August 5, 1939. The second session was relatively short, from September 21 to November 3,

1939. During that session, the committee and full Senate took no action on the USTB bill.

82. "Memorandum on Travel Bill H. R. 6884," n.a., n.d. (probably fall 1939). Box 134.

Notes to Chapter 5

1. "Americas' Travel Shows Sharp Rise," *NYT*, December 9, 1940, 27.
2. Presidential Proclamation No. 2382, January 13, 1940, 54 *Stat.* 2681–82.
3. Promoting closer relations with Latin America also had domestic political overtones. It was subtly subversive of the strident isolationism then popular with conservatives. Did their Fortress America ideology mean even opposition to the Monroe Doctrine and generally to closer relations with South America? And if doing *that* was okay by them, then why not with ——? The joint governmental and business efforts to promote tourism between the Americas was sometimes dubbed the "American Axis." "Air and Ship Lines to Promote Travel on 'American Axis,'" *WS*, September 29, 1940, B-6. This moniker quickly died out because that same month Germany, Italy, and Japan became allies and were known as the Axis Powers.
4. "Campaign Started to Extend Travel in Both Americas," *NYHT*, January 21, 1940, D11; "Accelerate 'Travel America Year'" (photo and caption), *WP*, January 20, 1940, 16. Pictured were D. Leo Dolan, head of the Canadian Travel Bureau, General William D. Leahy, governor of Puerto Rico, and Macnamee.
5. *CSM*, January 3, 1940, 4; *NYHT*, January 14, 1940, 34; *NYT*, January 20, 1940, 32. The *Times* article also demonstrated again the close link between travel promotion and federal PR. Robert Horton, the US Maritime Commission's PR director, was invited to address a conference Ickes convened of travel business and industry leaders. Horton assured them that inter-American sea travel was safe and that German submarine scares "proved to be entirely unfounded."
6. Charles M. Macko, "It's Travel America Year," *Barron's* 20:17 (April 22, 1940) 7–8.
7. The military draft started in the fall of 1940. It had the effect of increased nontourism travel by draftees and service men. This put some strain on available train and bus capacity for tourists. On the other hand, the travel industry saw other positive impacts of the draft. Some families traveled to meet up with husbands and sons in the service. Similarly, weekend passes meant that members of the armed forces visited nearby cities and spent money on lodging, food, and entertainment.
8. Walter R. Bottcher, AP, "Tourists on the Move after Delayed Start," *Baltimore Sun*, July 28, 1940, CS-2.
9. "National Park Records Show Gain in Travel," *CT*, October 16, 1940, 22.
10. Robert L. Perry, "Quiet Mexico Still Popular," *Detroit Free Press*, October 27, 1940, Part 1, 13.

11. Macnamee, "Travel Looks Ahead to Another Big Year" (guest column), *NYT*, December 29, 1940, XX-1.

12. USTB Organization Chart, October 15, 1940. Organization Charts 1934–1971, Box 1, Entry P23, National Park Service, RG 79.

13. 54 *Stat.* 652.

14. *Detroit Free Press*: "Travel Between Americas Built Up over Forty Years," March 31, 1940, Part 1, 15; "Tour Officials Ask Solidarity," September 7, 1940, Part 1, 12; "Railroads Adding Equipment," September 15, 1940, Part 1, 7.

15. *OB*: "Bureau Launches Poster Clearing House," 24 (April 1940) 2; "Travel Posters Listed in Bureau Publication," 26 (June 1940) 7.

16. "Material Available for Travel Displays," *Women's Wear Daily*, February 9, 1940, 7.

17. "Bureau Launches Poster Clearing House," *OB* 24 (April 1940) 2.

18. For example, "Mississippi," *OB* 12 (August 10, 1939) 4.

19. *OB*: "Florida," 21 (January 1940) 12; "Delaware," 22 (February 1940) 12; "Nebraska," 23 (March 1940) 12; "Arizona," 25 (May 1940) 12; "Virginia," 26 (June 1940) 12; "Maryland" and "Texas," 28 (September 1940) 12.

20. "Added Library Facilities," *Travel News* 2:23 (December 16, 1940) 8.

21. *OB*: 15 (October 10, 1939) 4; 12 (August 10, 1939) 2.

22. *OB*: 23 (March 1940) 12; 25 (May 1940) 12.

23. *Detroit Free Press*: "Publicity Push to Boom State," April 8, 1940, Part 1, 7; Robert L. Perry, "More Time Off Puts Travel Up," September 1, 1940, Part 1, 14. He also helped the Council of State Government's 1940 *Advertising by the States.*

24. Macnamee: "Summer of 1940 Promises Rush of Americans on Tour," *CSM*, June 7, 1940, V2; "Six Billion for Travel in '40," *Washington Times-Herald*, June 16, 1940, E-1, E-6.

25. "The Shape of Things to Come," *OB* 29 (October–November 1940) 7.

26. Herman Radolf, "Travel—Over American Trails," *Women's Wear Daily*, March 14, 1940, Section 2: "Summer Session," 24, 28.

27. "Travel," *Detroit Free Press*, October 20, 1940, Part 1, 13.

28. "Special Exhibit Devoted to the Grazing Service," *The Grazing Bulletin* 3:2 (March 1940) 29; "C.A.A. Exhibit to Go to San Francisco Fair," *WS*, June 1, 1940, A-2.

29. "New Quarters," *Travel News* 2:21 (November 16, 1940) 1–2.

30. "Notice of Suspension," *Travel and Recreation News Letter*, March 11, 1940. The last substantive issue was no. 12, dated February 20, 1940.

31. Meredith Mendelsohn, "Of Guidebooks to the American Dream," *NYT*, January 21, 2018, AR-18.

32. The 2018 movie *Green Book* was based on the use of a later edition of the guidebook by an African American pianist and his white driver during a concert tour of the South in 1962.

33. "Guide for Negro Motorists," 27 *OB* (July–August 1940) 12.

34. McDowell, "Travel Notes," *Travel and Recreation News Letter* 10 (January 20, 1940) 8.

35. Ibid., 11 (February 5, 1940) 9; 12 (February 20, 1940) 6.

36. August 15 was Thursday, so it was a relatively lower attendance day compared to weekends.

37. "Committee Discusses 'Negro Day' at Fair," *Norfolk [VA] New Journal and Guide*, February 24, 1940, 5.

38. "World's Fair Plans Sepia Beauty Contest," *Philadelphia Tribune*, May 9, 1940, 14.

39. *CR* 86:13 (February 22, 1940) 905–06; 86:18 (December 18, 1940) 6911–12.

40. *CR* 86:8 (July 2, 1940) 9195.

41. Ibid., 9196.

42. "Dividend Due!" *OB* 29 (October–November 1940) 7, emphasis added.

43. Letter from Roger Williamson, Assistant Clerk, Senate Commerce Committee, to Breckinridge Long, Assistant Secretary of State, July 10, 1940. File: H.R. 6884, 76th Congress, Records of the Committee on Commerce and Related Committees 1816–1968, US Senate Records (RG 46), CLA.

44. Letter from Ickes to FDR, March 7, 1940. Box 134.

45. Memo from NPS Associate Director Arthur Demaray to Mr. Dunlap, March 12, 1940. Box 134.

46. Letter from Roger Williamson, Assistant Clerk, Senate Commerce Committee, to Senator Ernest W. Gibson, February 21, 1940. File: H.R. 6884, 76th Congress, Files of the Senate Commerce Committee.

47. For Ickes, passing this bill was a major personal effort. For BOB, it was a relatively minor and routine matter. Notwithstanding formal correspondence signed by FDR (see next note), BOB director Harold Smith does not even mention dealing with this subject during March 1940. Summaries of daily activities (Folder 11, Box 1) and meetings with the president (Folder 2, Box 3), Papers of Harold D. Smith, FDR Library.

48. Letter from FDR to Ickes, March 22, 1940. Box 134. First Assistant Secretary Ebert K. Burlew quickly followed up by talking to assistant BOB director F. J. Bailey (no relation to the senator), head of the legislative clearance office. He wanted to know precisely what changes were absolutely necessary to get BOB's signoff on the bill. Bailey said he would talk it over with his (understandably punctilious) staff "to see whether they might not relax a little bit from the requirements of the letter." Letter from Burlew to Macnamee, March 30, 1940. Box 134.

49. He was also openly anti-Semitic and anti-Italian.

50. Letter from Macnamee to Bilbo, March 2, 1940. Box 134.

51. *CR* 86:4 (April 10, 1940) 4268–69.

52. Letter from Roger Williamson, Assistant Clerk, Senate Commerce Committee, to C. Pitman Baker, Jr., Hotel Colonial, Philadelphia, March 20, 1940. File: H.R. 6884.

53. Letter from Ickes to Barkley, April 10, 1940. Box 134.

54. *CR* 86:4 (April 10, 1940) 4269.

55. Yet *again*, the State Department informed the Senate of its concerns that some people (read: German Jews) might take advantage of any new policies that would make it easier to obtain tourist visas. In March 1940, Assistant Secretary Breckenridge Long wrote that "under the present circumstances a large proportion of such persons would actually desire to emigrate to this country. It would then become the duty of our consular establishment to refuse visas to such persons and in many cases it would be exceedingly difficult accurately to determine their status" (i.e. motivations and intent). Letter from Long to Senator Bailey, March 22, 1940. File: H.R. 6884. Between its previous letter of February 1939 (chap. 4) and March 1940, Germany had invaded Poland and began deporting the remaining Jews in Germany to ghettos and concentration camps in Poland and Czechoslovakia.

56. Ickes Diary (unpublished version), April 26, 1940, p. 4351.

57. Letter from Ickes to Barkley, June 17, 1940. Box 134.

58. *CR* 86:8 (June 22, 1940) 8985–86.

59. *CR* 86:9 (July 11, 1940) 9494 (Senate) and 9540–41 (House).

60. 54 *Stat.* 773–74.

61. "President Signs Travel Bill," *Railway Age* 109:4 (July 27, 1940) 156; Dunlap, "Travel Bureau's Promotion Work Wins Recognition from Congress," *NYHT*, September 8, 1940, D11. USTB's *Official Bulletin* had printed a celebratory statement by Ickes upon the signing of the bill on the front page of its July–August issue (no. 27). Obsequiously, Macnamee sent a copy of the *Herald Tribune* article to Ickes and, in a cover letter, further praised Ickes's "striking statement to the travel industry." Memo from Macnamee to Ickes, September 13, 1940. File: Articles—General January 4, 1940–January 3, 1941, Box 100, Ickes Papers. Despite his protestations that he was a curmudgeon and thrived on media criticism, Ickes meticulously saved clippings of positive coverage (as well as negative) in his office files.

62. Memo from NPS Acting Director Hillory A. Tolson to Washington Office and All Field Offices, October 4, 1940. Box 134.

Notes to Chapter 6

1. This chapter covers events in 1941 through the Japanese attack on Pearl Harbor on December 7.

2. FDR had proclaimed that the United States was in a *limited* state of national emergency in September 1939, a week after the Nazi invasion of Poland. The invasion had triggered declarations of war by France and the United Kingdom.

3. "No Money for National Park Service's Travel Bureau," *Railway Age* 110:19 (May 10, 1941) 814.

4. "Uncle Sam's Travel Bureau" (editorial), *NYT*, May 22, 1941, 20.

5. He, of course, had very strong parochial interests in other activities of Interior, such as irrigation canals, dams, mining, grazing, and Indian reservations.

6. 55 *Stat.* 352.

7. This was a good demonstration of Parkinson's law of triviality, that the length of discussion of a subject is in inverse proportion to the amount of money being discussed as well as in direct proportion to the relatability and tangibility of the subject to the decision makers.

8. 55 *Stat.* 277.

9. 55 *Stat.* 561.

10. "Jay Wingate New Supervisor Eastern Branch," *OB* 2:3 (March–April 1941) 6.

11. US Civil Service Commission, *Official Register of the United States, 1941*, pp. 80, 89, 94. This was a congressionally mandated annual report and was dated May 1, 1941. The three were also listed in the *Official Congressional Directory* for the 77th Cong., 1st sess., 2nd ed. (updated through April 23, 1941), p. 340.

12. *OB* 2:2 (February 1941) 5; *Travel News* 3:3 (February 1, 1941) 7; *Eastern Travel Today* 4 (June 15, 1941) 7; *Travel West* 3:13 (July 1, 1941) 7.

13. Memo [n.t.], from Hillory A. Tolson, NPS Acting Associate Director, to A. E. Demaray, NPS Acting Director, July 17, 1941. Box 134.

14. "U.S. Travel Bureau Has Daily Film Show," *NYHT*, January 19, 1941, D11; "A Travel Miscellany: Tourist Bureau Movies," *NYT*, February 2, 1941, XX-8.

15. Dunlap, "Value in 'Health, Wealth and Unity' Cited by Federal Travel Officials," *NYHT*, February 2, 1941, E11.

16. "Films are Loaned by Travel Bureau," *Eastern Travel Today* 4 (June 15, 1941) 14.

17. "Film of Latin America to be Shown Tonight," *WP*, June 13, 1941, 17.

18. "Free Film Reviews," *Movie Makers* 17:2 (February 1942) 52.

19. Robert H. Wall, "People You Should Know," *Travel West* 3:16 (August 16, 1941) 7.

20. "Radio Broadcasts," *Boston Globe*, July 3, 1941, 20.

21. "Sources of Visual Aids for Instructional Use in Schools," Pamphlet No. 80, 1941, US Office of Education, 5, 22, 27; Marie K. Pidgeon, "America in the Graphic Arts," *High Points* 23:3 (March 1941) 35 (published by the Board of Education of the City of New York).

22. Horton S. Allen, Jr., "Allen talking—" (column), *Eastern Travel Today* 4 (June 15, 1941) 11.

23. "Wall Street Scene" (column), *Wall Street Journal*: January 17, 1941, 4; January 21, 1941, 4.

24. "Report on Sports Areas," *NYHT*, February 2, 1941, F10.

25. "Two Vacations a Year Seen as Major Asset in Travel Promotion," *OB* 2:2 (February 1941) 8.

26. "Neighbors at Show," *NYT*, March 16, 1941, AG25; George H. Copeland, "A Program for Travel," *NYT*, October 19, 1941, XX-1; "Travel Bureau Chief Arrives," *NYHT*, December 4, 1941, 27.

27. Macnamee: "Travel in the United States Urged as an Aid to Unity and Patriotism," *NYHT*, April 20, 1941, I2; "Travel Adds Six Billion to 1941 Income," *Detroit Free Press*, May 4, 1941, 1–10; "17 Million Cars in Use for Holiday," *Detroit Free Press*, June 1, 1941, 1–8; "Outlook Big for Travel in United States," *CSM*, June 6, 1941, V1.

28. For example: "Travel Bureau Survey Shows Parks Have Monetary Value to States," *Amarillo [TX] Daily News*, February 7, 1941, 10; Dunlap, "1941, All-American Year, Promises Record Travel," *NYHT*, April 20, 1941, I2; John Markland, "Two-Way Travel over Holiday," *NYT*, June 29, 1941, XX-1; "Standard Tourist Cards Proposed," *San Antonio [TX] Light*, October 1, 1941, 4-A; "Pleasure Trips Urged as Morale Builders in Defense Effort," *WS*, October 13, 1941, A-11; "More Americans Planning Wintertime Vacation Trips," *WP*, November 16, 1941, L11.

29. "Random Notes for Travelers: Vacation Store House," *NYT*, June 8, 1941, XX-11.

30. USTB published follow-up monthly supplements with updated information in February, April, May, and June, 1941. These mimeographed publications were located in the federal documents collection at the Wisconsin Historical Society library. The travel section of the Sunday *Herald Tribune* published an extensive two-page list of events that USTB had prepared and circulated: "Events and Their Dates to Attract the Vacationist," April 20, 1941, I12–13.

31. "U.S. Travel Bureau Maintains Lists of Holiday Areas," *NYT*, June 8, 1941, XX-11.

32. "Branch Office Publications Expand," *OB* 2:3 (March–April) 6.

33. It is possible that the withdrawal of WPA staffing and assistance from the New York office at the end of FY 1941 may have been a factor in the interruption. The issues in the spring of 1941 were numbered (1–4), but without volume identification. The October and November issues (5–6) were listed as part of volume 1.

34. AP, "Promotion Unit Urges Nation to Visit West," *Salt Lake [City, UT] Tribune*, April 5, 1941, 3.

35. "Should Holiday Dates Be Changed?," *OB* 2:1 (January 1941) 4; "Ayes and Nays on Changing of Holiday Dates," *OB* 2:4 (May–June 1941) 4; "Resort Men Propose to have All Holidays Fall on Mondays," *WS*, July 7, 1941, B-9.

36. Futrelle, "Traveler's Notebook" (column), *WS*, April 13, 1941, B-4.

37. M. W. Daub, "Luggage Industry Develops in Step with Travel," *OB* 2:3 (March–April 1941) 8.

38. Horton S. Allen, Jr., "Allen talking—" (column), *Eastern Travel Today* 4 (June 15, 1941) 11.

39. Indicating the disruption caused by WWII, the next edition of the *Green Book* was issued in 1947.

40. "1941 Edition of Negro Motorist Guide," *OB* 2:4 (May–June 1941) 12. The only content after this article was a list of recently issued travel booklets. That hotels were assumed to be restricted to white guests was common throughout the United States, not just the South.

41. McDowell, "State Park Recreational Areas," *Eastern Travel Today* 1 (March 15, 1941) 7.

42. McDowell, "Conference of Negro Recreation Workers," *Eastern Travel Today* 2 (April 15, 1941) 8.

43. McDowell, "Negro Day Pilgrimage to Roanoke Island, N.C.," *Eastern Travel Today* 4 (June 15, 1941) 8. A different story in that issue was about tourism to the Outer Banks of North Carolina. It contained this observation about one of the islands: "Its population is less than 100, including one Negro family." "North Carolina's Shangri-La," ibid., 5.

44. James A. Jackson, "The Negro Also Travels," *Eastern Travel Today* 3 (May 15, 1941) 8.

45. "Travel America" (column), *Baltimore Afro-American*, August 30, 1941, 6.

46. ANP, "Travel Bureau Is Eliminated," *Atlanta Daily World*, December 1, 1941, 1. A week later (the day before Pearl Harbor), the national weekly edition of the *Chicago Defender* ran the same item: "Travel Bureau at N.Y. Is Dissolved," December 6, 1941, 12.

47. *Atlanta Daily World*, ibid.

48. USTB's outreach to African Americans was remembered after McDowell left the agency. A 1943 article on hotels for African Americans cited statistics from USTB's 1941 directory. "N.Y. Leads Cities with Race Hotels," *Baltimore Afro-American*, August 28, 1943, 6. When the *Green Book* resumed publication after the war, the news coverage reminded readers of USTB's association with it. "Motorist Guide Book Returns after 4 Yrs," *Philadelphia Tribune*, April 13, 1946, 16. After leaving USTB, McDowell continued working in the travel business. In 1942, he was working with Charles H. Bailey and Revella Hotels, presumably a hotel chain catering to African Americans. Ida M. Smith, "Travel America" [column], *Baltimore Afro-American*, July 25, 1942, 8. After that, he was associated with the Fugazy Travel Bureau. "Socially Speaking" [column], *New York Amsterdam News*, November 10, 1951, 19. A reporter who interviewed him in 1948 wrote that "Mr. McDowell is a man with a dream, that one day Negroes will go anywhere using the best means of transportation and will be freely accepted everywhere." Lillian Scott, "Name Your Destination: McDowell Gets You There," *Chicago Defender* (national weekly edition), June 12, 1948, 13. He died in 1953. "McDowell, 69, Travel Agent, Dies," ibid., December 26, 1953, 4.

49. P. V. G. Mitchell, vice president, United States Lines, "Merchant Marine Develops New Travel Fields," *OB* 2:3 (March–April 1941) 1.

50. "Use of American Spas Urged by Macnamee," *Eastern Travel Today* 1 (March 15, 1941) 2.

51. "Travel Strengthens America: Program for 1941," *OB* 2:1 (January 1941) 7. A matchbook company, in cooperation with USTB, designed a match cover with Uncle Sam pointing to a US map and saying, "Come on, folks . . . It's all yours" (ibid., 10).

52. *Travel News* 3:2 (January 16, 1941) 1–3.

53. Bossemeyer, *Travel News* 3:2 (January 1, 1941) 1–2.

54. J. Leslie Kincaid, president, American Hotels Corporation, "Travel and the American Way of Life," *OB* 2:2 (February 1941) 1.

55. Dunlap, "Value in 'Health, Wealth and Unity' Cited by Federal Travel Officials," *NYHT*, February 2, 1941, E11.

56. Bossemeyer, "National Defense and the Travel Industry," *Travel West* 3:8 (April 16, 1941) 1–2. The important detail of being categorized as a defense industry conveys how much the national economy was subject to the command and control of federal agencies managing the defense mobilization efforts *before* Pearl Harbor. A similar struggle occurred within the executive branch with many agencies seeking to be defined as defense related and, later, war agencies (Lee 2018, 189).

57. Macnamee, "Emergency Spurs Travel with a Patriotic Purpose," *OB* 2:4 (May–June 1941) 2.

58. Lee P. Hart, "The Travel Outlook" (column), *OB* 2:4 (May–June 1941) 7.

59. Dunlap, "1941, All-American Year, Promises Travel Record," *NYHT*, April 20, 1941, I2.

60. Leavitt F. Morris, "Boom Travel Season Seen in Americas," *CSM*, June 6, 1941, Section 2, 1.

61. "Temporary Suspension of Passenger Service," *Travel West* 3:9 (May 1, 1941) 4.

62. "Asilomar Taken Over by the Government," ibid., 10.

63. "Army Traffic," *Travel West* 3:10 (May 16, 1941) 20.

64. "Conserve Airline Space," *Eastern Travel Today* 1:5 (October 1941) 11. In those days, reservations could be held for long periods because *paying* for the ticket could only be done in person and either by check or cash. (Credit cards had not yet been invented.) Payment was sometimes done at a travel agency or if the airline had a downtown ticket office. However, sometimes the ticket was not paid for until the passenger checked in at the airport. Airlines did not have standard policies for how long they would honor a reservation before canceling it, nor was it common to have stand-by passengers waiting at the gate in case of no-shows.

65. "Latest Control Tower Devices Insure Safety of Air Travelers," *OB* 3:1 (July–August 1941) 9; "'Detector' Rides Rails for Safety," *Eastern Travel Today* 1:6 (November 1941) 8.

66. Bossemeyer, "Travel Is Important Factor in U.S. Economic System," *Oakland [CA] Tribune*, May 11, 1941, 6-C.

67. Macnamee, "The Trend in Travel," *Travel West* 3:11 (June 1, 1941) 1–2.

68. Conquered from Spain in the Spanish American War, by now the islands were largely self-governing and autonomous. The office of commissioner was partly ceremonial.

69. FDR wanted to create a cabinet-level department of public welfare, but Congress would not approve it. FSA was a de facto department without the formal legal status. President Eisenhower reorganized it into the Department of Health, Education, and Welfare.

70. Mary Hornaday, "Washington Side-Glances," *CSM*, June 13, 1941, 10.

71. McNutt, "Travel to Keep Americans Fit for Patriotic Duty Urged by Defense Recreation Coordinator McNutt," *OB* 3:1 (July–August 1941) 1.

72. "McNutt Stresses Value of Travel in Nation's Welfare," *NYHT*, August 24, 1941, D11; "Tourism's Role in National Defense," *NYT*, August 31, 1941, XX-5; "Festival Held Defense Aid," *LAT*, September 23, 1941, A1.

73. June 8, 1941: "17,000,000 Autos to Carry Hundreds of Thousands of Persons on Short Holidays to Lakes and Mountains this Summer," *WP*, L7; "17 Million Automobiles Available to Tourists," *NYHT*, D14.

74. In early November, FDR enhanced Ickes's powers to control gasoline supplies by expanding his portfolio to include other forms of energy. Ickes's new title was Solid Fuels Coordinator for National Defense (Roosevelt 1969, 1941: 470–72).

75. "Gasless Sundays in the West," *Travel West* 3:13 (July 1, 1941) 12.

76. Lee P. Hart, "The Travel Outlook" (column), *OB* 3:1 (July–August 1941) 4.

77. "Motorists Big Aid to Nation: Keystone Auto Club Takes Issue with Sec. Ickes' Gas Shortage Threat," *Chester [PA] Times*, June 30, 1941, 18.

78. "The Itching Foot" (editorial), *NYT*, August 29, 1941, 16.

79. "OPM Curtails Nonmilitary Use of Rubber," *WP*, June 20, 1941, 6.

80. "Retreading of Tires Expected to Reach New Record This Year," *Wall Street Journal*, July 12, 1941, 7.

81. "Tire Sales Hit Record on Fears of Shortage," *NYT*, August 5, 1941, 28.

82. Memos from Horton S. Allen Jr., Acting Supervisor, New York Branch Office, to USTB chief, August 20 and September 5, 1941. Box 135.

83. "President Hails Recreation Drive," *NYT*, September 30, 1941, 26.

84. "Recreation Vitally Needed by Nation Now, McNutt Tells Convention in Baltimore," *WP*, September 30, 1941, 8.

85. "The President Urges Travel Development to Maintain National Income and Morale," *OB* 3:2 (November 1941) 7. The wording of the headline by *OB*'s editor (a USTB employee) was grossly misleading, self-serving, and a far stretch from what FDR actually said.

86. "Travel Strengthens America," *Travel West* 3:19 (October 16, 1941) 1.

87. Ibid., 10–12.

88. "Check on Inflation Urged at Conference," *Boston Globe*, November 15, 1941, 11.

89. "Travel Officials Pledge Aid to National Morale," *NYHT*, October 17, 1941, 15.

90. "Travel Agencies of Continent to Merge Efforts," *NYHT*, October 15, 1941, 21.

91. "Conference Speakers See Travel as Aid in Defense Morale," *WS*, October 15, 1941, A-11. A similar quote appeared in George H. Copeland, "A Program for Travel," *NYT*, October 19, 1941, XX-1.

92. *OB* 3:2 (November 1941) 11.

93. A somewhat parallel development occurred at the beginning of 2009, when the Great Recession was having major negative effects on the economy. Many conferences scheduled for Las Vegas were canceled. Less than a month after taking office, President Obama explicitly discouraged companies that were receiving federal economic stimulus funds from taking "a trip to Las Vegas or go down to the Super Bowl on the taxpayer's dime." It was apparently an off-the-cuff and ad-libbed comment. The travel industry in general and Las Vegas in particular howled in protest. Sholnn Freeman and Michael S. Rosenwald, "Travel Industry: This Is No Time to Check Out," *WP*, February 13, 2009, D1; Steve Friess, "Las Vegas Sags as Conventions Cancel," *NYT*, February 15, 2009, 28.

Notes to Chapter 7

1. This chapter also includes events in late 1941 after the Japanese attack on Pearl Harbor on December 7.

2. "Travel West Monthly," *Travel West* 3:24 (December 16, 1941) 3. The date on the photo cover is December 15, but the date on the first page of the text is December 16.

3. Macnamee, "Travel and Morale," *WS*, December 14, 1941, D-14.

4. Macnamee, "Travel's War Role," *NYT*, December 21, 1941, XX-3.

5. Ibid., XX-7.

6. "Government to Encourage Travel during the War," *Boston Globe*, December 28, 1941, A15.

7. In July 1941, FDR signed an executive order creating the Economic Defense Board. On December 17, he signed another executive order renaming it the Board on Economic Warfare.

8. Dunlap, "1914–'18 Toll of Stamina Cited as Proof of Vacation Travel Need," *NYHT*, December 21, 1941, D10.

9. "Travel Talk," *Travel West* 4:1 (January 1942) 4–5.

10. "Montanans Hear of Effort Needed to Insure Victory," *Helena [MT] Independent*, April 11, 1942, 1–2.

11. January 25, 1942: "Veteran Ticket Agent Knows His Railroad," *WP*, L11; "Interior Sec. Ickes Encourages Travel," *Boston Globe*, C18.

12. Frederic Babcock, "Tourist Travel Recovers from December Panic," *CT*, January 14, 1942, 19.

13. Diana Rice, "Notes for the Traveler: Parks Await Throngs" (column), *NYT*, January 18, 1942, XX-4; Morris, "Tourabouts: Parks to Open," *CSM*, January 27, 1942, 11.

14. "National Parks Will Not Close," *Detroit Free Press*, February 1, 1942, 1–7.

15. For example, *NYHT*'s Dunlap wrote three articles referencing USTB in two days. January 4, 1942: "Health and Morale Linked to Maintenance of Vacation Travel," D10; "Ample Service Predicted for Winter Travel," H4. January 5, 1942: "Tourist Trade Seen Resisting Impact of War," 22. A few weeks later, he discussed it again: "Floral Regions of South are Famed as Early Spring Vacation Objective," February 1, 1942, D10.

16. AP, "Lack of Tires May Cut Down on S.D. [South Dakota] Tourists This Year," *[Huron, SD] Evening Huronite*, January 10, 1942, 7; "Experts See Busy Summer for New Hampshire Resorts," *Portsmouth [NH] Herald*, January 16, 1942, 9; "Resort Prospects" (Letter to the Editor), *[Benton Harbor, MI] News-Palladium*, January 28, 1942, 2.

17. "Text of Secretary Ickes' Press Release on Wartime Travel Policy," January 17, 1942, *OB* 3:3 (January–February 1942) 3. The press release from Interior's Information Service is in the USTB folders at the National Archives. Document control number P.N. 173106, Box 134.

18. "A Little Pleasure Still Room for Riding for Motorists," *Cullman [AL] Banner*, January 29, 1942, 3; "No Priorities on Travel as Government Okehs [sic] Trips to Keep America Fit," *Long Beach [CA] Independent*, February 1, 1942, 13; Morris, "America's Skies and Trees: Builders of Strength," *CSM*, February 27, 1942, 13; "Agencies Predict Brisk Wartime Travel to E.P. [El Paso]," *El Paso [TX] Herald-Post*, March 4, 1942, 3.

19. "Ickes Favors Continued Civilian Pleasure Travel," *Hotel World-Review* 122:4 (January 24, 1942) 7; "Tourist Business Goes On" (editorial), *Advertising Age* 13:4 (January 26, 1942) 12; "Flash: Ickes Reveals Governmental Policy on Travel," *The Travel Agent* 13:1 (January 1942) 17.

20. "Ickes Approves Civilian Travel as Morale Aid," *WS*, January 18, 1942, A-18; United Press, "Ickes Urges Vacation Travel," *NYT*, January 20, 1942, 17; "Civilian Travel for Relaxation Urged by Ickes," *NYHT*, January 20, 1942, 22.

21. Editorials: "Travel for Morale," *Anniston [AL] Star*, January 21, 1942, 4; "See America First," *HC*, January 27, 1942, 10 (reprint of "Mr. Ickes Urges Travel," *Worcester [MA] Telegram*, January 22, 1942). Ickes's clippings collection contains more than a dozen other similarly themed editorials published in small newspapers. Boxes OV 55–56, Ickes Papers.

22. Based on the scrapbooks of news coverage in his papers at the Library of Congress, he evidently subscribed to a national newspaper-clipping service unrelated

to OGR's clipping office. These clippings probably arrived at his office in a delayed fashion, and, even then, he may not have routinely looked them over.

23. Editorials: "Travel, but Not as Usual," *Toledo [OH] Blade*, January 23, 1942; "Ickes Favors Vacation," *Lamar [CO] News*, January 28, 1942. Box OV 55, Ickes Papers. "Wartime Vacations," *Asheville [NC] Times*, March 11, 1942. Box OV 56, Ickes Papers. Ickes's January press release was not enough for Florida governor Spessard Holland. Two weeks later, he asked governors of other states with major tourism destinations to join him in asking FDR to issue a presidential statement encouraging continued travel and tourism during the war. AP, "Florida Governor Urges Action to Spur Travel," *Helena [MT] Independent*, February 2, 1942, 1. Quickly, another newspaper condemned his parochialism. "Selfish Florida" (editorial), *Pawtucket [RI] Times*, February 3, 1942. Box OV 55, Ickes Papers. For the next few months, Holland continued complaining about Ickes and gas rationing. AP, "Gasoline Order Called Unfair by Governor," *WP*, March 16, 1942, 12; "No Reason for Tourist Alarm" (editorial), *Tampa [FL] Tribune*, March 18, 1942, Box OV 56, Ickes Papers; AP, "Ration Amid Plenty," *CT*, May 17, 1942, 14.

24. *OB* 3:3 (January–February 1942) 1.

25. Ibid., 2.

26. "Petroleum Coordinator Urges Voluntary Gasoline Conservation," ibid., 20.

27. "Dude-Ranch Activities Expected to Expand Widely Next Summer," *NYHT*, March 15, 1942, E9.

28. *CR* 88:3 (March 26, 1942) 3016.

29. "Harold's Whifflebat Bulletin" (editorial), *Macon [GA] Telegraph and News*, March 8, 1942. Box OV 56, Ickes Papers.

30. "Ickes as Usual" (editorial), *Evansville [IN] Courier-Press*, March 8, 1942. Box OV 56, Ickes Papers.

31. "An Example of Non-Essential Federal Spending" (editorial), *Waterloo [IA] Daily Courier*, March 9, 1942, 4.

32. "Mr. Ickes' Travelog" (editorial), *Manchester [NH] Union*, March 9, 1942. Box OV 56, Ickes Papers.

33. "Confusion Worse Confounded" (editorial), *Gastonia [NC] Daily Gazette*, March 11, 1942, 4.

34. Lewis T. Nordyke, "'Doc' Ickes Prescribes Planned Vacation for Tense Americans," *Amarillo [TX] Daily News*, March 11, 1942, 6. Ickes's clippings contain another half dozen similarly themed editorials published that week.

35. Ickes Diary (unpublished version), March 15, 1941, pp. 6434–35. Ickes dictated this diary entry only two days after his meeting with Macnamee, so the credibility of his version is strong (unlike post hoc memoirs), even if he may have been presenting a relatively heroic narrative of his actions.

36. "Double Role Gets Ickes in Doghouse," *Salt Lake [City, UT] Tribune*, April 5, 1942, A-9. Editorials: "War-Time Travel Policy," *Albuquerque [NM] Journal*,

March 15, 1942, 10; A. Q. Miller, "Observations," *Belleville [KS] Telescope*, March 19, 1942, B-7. Column: "Iffy the Dopester," *Detroit Free Press*, March 25, 1942, 26.

37. Ickes Diary (unpublished version), March 15, 1941, pp. 6434–35.

38. *Official Congressional Directory*, 77th Cong., 2nd session, 1st ed. (December 19, 1941), p. 349.

39. "Travel Talk," *Travel West* 4:1 (January 1942) 22. This ended up being the last issue of *Travel West*.

40. "Filling War Role, United States Travel Bureau Moves Washington Office to New Interior Bldg.," *OB* 3:3 (January–February 1942) 19.

41. "U.S. Travel Bureau Closes Office Here," *NYT*, December 28, 1941, XX-5. Exactly when they closed is unclear. Articles on the Sunday travel pages in February and March continued to refer to the offices in New York and San Francisco as sources of information for travelers: "Tips to Tourists," *CT*, February 15, 1942, F4; Dunlap, "National and State Parks Provide Week-End Vacation Trip Facilities," *NYHT*, March 22, 1942, D9.

42. "Filling War Role, United States Travel Bureau Moves Washington Office to New Interior Bldg.," *OB* 3:3 (January–February 1942) 19.

43. Ray Tucker, "National Whirligig: News behind the News" (syndicated column), *Muscatine [IA] Journal and News-Tribune*, April 2, 1942, 4.

44. United Press, "Federal Officials Squelch Rumors of Vacation Ban," *Racine [WI] Journal-Times*, April 13, 1942, 18.

45. Wingate letter to Mr. Cass, March 27, 1942; Wingate memo to Miss Mathilda C. Heuser, USTB Chief Clerk, April 8, 1942; Heuser memo to Wingate, April 17, 1942; Wingate memo to Macnamee, April 17, 1942. Box 135.

46. Letter from Macnamee to Herbert Kahler, NPS, December 28, 1942. Box 134.

47. *Official Congressional Directory*, 77th Cong., 2nd session, 2nd ed. (May 26, 1942), p. 351.

48. Morris, "Travel and Vacation Facilities in War Supported at Atlantic City Session," *CSM*, February 16, 1942, 19.

49. "Keystone Club Board Backs 9:30 Opening for D.C. Schools," *WS*, February 18, 1942, B-2.

50. Rice, "Notes for the Traveler" (column), *NYT*, February 15, 1942, XX-2.

51. "Policy to Advance Travel Adopted," *NYT*, February 28, 1942, 18.

52. United Press, "Federal Officials Squelch Rumors of Vacation Ban," *Racine [WI] Journal-Times*, April 13, 1942, 18.

53. Rice, "Notes for the Traveler: Dude Ranch Plans" (column), *NYT*, April 19, 1942, D9. NPS's Director Newton Drury was quite agitated when he heard about this. He wired Wingate, the New York office supervisor, asking why the office was still engaging in PR when it was supposed to be shutting down. Wingate replied defensively that he did little other than to make room arrangements with the hotel manager. Quoted in a memo from Wingate to Macnamee, April 17, 1942. Box 135.

54. Marshall Sprague, "The West Will Do Handsomely by Its Visitors This Summer," *NYT*, May 10, 1942, XX-1.

55. Santa Fe Railroad: "What About Travel to California Today?" (ad), *NYT*, January 25, 1942, XX-6. Southern Pacific Railroad: "The Victory Trains Are Rolling!" (ad), *LAT*, January 28, 1942, 7; "Thumbs Up!" (ad), *Tucson [AZ] Daily Citizen*, February 16, 1942, 12.

56. All-Year Club of Southern California, "Henry Is 10 Years Younger!—Thanks to His Civilian Furlough" (ad), *CT*, May 10, 1942, H7.

57. Hotels of Southern California, "We're Doing Our Part . . ." (ad): *Amarillo [TX] Globe*, July 23, 1942, 11; *Tucson [AZ] Daily Citizen*, July 28, 1942, 2; *[St. Louis, MO] Sporting News*, August 20, 1942, 10.

58. *NYT*, March 1, 1942, XX-5; *CSM*, April 11, 1942, WM-12.

59. "Travelers Please Note," *NYHT*, April 24, 1942, 13.

60. *CR* 88:8 (February 24, 1942) A679–80.

61. *CR* 88:3 (March 27, 1942) 3086–89. Rees returned to the subject in the fall during a debate about federal wartime employment. *CR* 88:6 (October 15, 1942) 8233.

62. *CR* 88:3 (May 13, 1942) 4161.

63. "Senate and House Battle over Farm Parity and Economy," *WP*, June 26, 1942, 14.

64. 56 *Stat.* 553.

65. AP, "U.S. Travel Board Asks Own End—Employs 3," *NYHT*, December 10, 1942, 12. Termination letters from Hillory Tolson, Acting NPS Director, to Heuser and Henni, February 17, 1943. Box 134.

66. Memo from Macnamee to NPS Director, October 22, 1942; Macnamee, untitled draft of summary of USTB activities in 1942 (probably for a more comprehensive NPS report on its wartime activities), November 4, 1942. Box 135.

67. "Government Travel Bureau Asks to be Abolished," *Railway Age*, 113:25 (December 19, 1942) 1011.

68. Macnamee, untitled draft of summary of USTB activities in 1942, November 4, 1942. Box 135.

69. NEA [Newspaper Enterprise Association, a Scripps news service], "Vacationists Using Over-Laden Railroads Must Remain Prepared to Say: 'After You, Soldier,'" *Laredo [TX] Times*, July 26, 1942, 8, emphasis added.

70. "Possibilities Considered," *Cumberland [MD] News*, July 21, 1942, 9. The article stated that he would be in the area of two weeks, which seems unlikely, a reminder that Congress had zeroed out USTB's travel budget for FY 1943. It is probable that Macnamee was able to take the trip because the B&O Railroad paid for his train ticket, and perhaps the local host organization covered his hotel costs.

71. "Resort Has Possibilities," *Cumberland [MD] News*, July 24, 1942, 13, 10.

72. AP, "U.S. Travel Bureau Chief 'Sold' on Garrett County," *Baltimore Sun*, July 25, 1942, 19.

73. Memo from Macnamee to NPS Director, October 22, 1942. Box 135.

74. Jerry Kluttz, "U.S. Travel Bureau Gets Ax for Duration" (Federal Diary daily column), *WP*, December 9, 1942, B1.

75. AP, "U.S. Travel Bureau Director Urges Ending Own Job," *WP*, December 10, 1942, 2.

76. December 10, 1942: "U.S. Travel Bureau Fades," *NYT*, 46; "Federal Body Suggests Its Own Discontinuance," *Baltimore Sun*, 6; "Head of U.S. Travel Bureau Seeks to Have Agency Closed," *WS*, A-3; "Bureau Recommends Own Discontinuance," *HC*, 1.

77. *Galveston [TX] Daily News*, December 10, 1942, 2.

78. *Salt Lake [City, UT] Tribune*, December 10, 1942, 7.

79. *[Jefferson City, MO] Daily Capital News*, December 10, 1942, 10.

80. December 11, 1942: AP, "Macnamee to Get Wish of Travel Bureau's End," *CSM*, 8; "Officials Grant Request to Close Travel Agency," *WS*, A-12. The anonymous BOB official was almost surely Assistant Director Wayne Coy. As a former reporter he was comfortable dealing with journalists. Lee 2018, chap. 10. Also, director Harold Smith's call log for December 9–10 does not list any contacts with an AP reporter. June–December 1942, Daily Record, Folder 6, Box 2, Smith Papers, FDR Library.

81. *Berkshire County [Pittsfield, MA] Eagle*, December 30, 1942, 10.

82. "Hail, MacNamee [sic]," *Charleston [WV] Daily Mail*, January 2, 1943, 4.

83. AP, "Spoon River Sam Says . . . ." (opinion column), *[Fayetteville] Northwest Arkansas Times*, January 19, 1943, 2; George S. Benson, "Looking Ahead" (editorial column), *Blytheville [AR] Courier News*, January 8, 1943, 3.

84. "Bureau Buster" (editorial), *[Massillon, OH] Evening Independent*, December 18, 1942, 4. This was a syndicated editorial and appeared in other small papers.

85. "It's About Time" (editorial), *El Paso [TX] Herald-Post*, December 26, 1942, 4.

86. Memo from Macnamee to NPS Director, October 22, 1942. Box 135.

87. After the public announcement and while winding down USTB, Macnamee apparently reverted to working in his pre-USTB position in the secretary's office. Filling in for the special assistant to the secretary, he handled some correspondence for Ickes. Letter from Macnamee to Miss Suzanne T. Gouriou, December 14, 1942. File: Invitations December 15, 1942–January 1943, Box 201, Ickes Papers.

88. Termination letters from Hillory Tolson, Acting NPS Director, to USTB employees Heuser and Henni, February 17, 1943. Box 134. Until her death in 1960, Heuser continued working in other offices at Interior, lastly at the Board of Geographic Names. "Miss Heuser, Interior Dept. Official, Dies" (obit.), *WP*, November 13, 1960, B15. Henni had graduated from the University of Omaha's Law School, but it is unclear if she was a practicing lawyer or engaged in any legal work while at USTB. After leaving USTB in 1943, she married, but her husband died in 1948. She was a widow until her death in 1982. In her will, she bequeathed

$116,000 to the Omaha Public Library for its South Omaha branch, which she had patronized while growing up. "News of Nebraska Libraries," *Overtones: Newsletter from the Nebraska Library Commission* 11:1–2 (January/February 1984) 6.

89. Memo from Macnamee to NPS Associate Director, December 26, 1942. Box 134.

90. Letter from A. E. Demaray, NPS Associate Director, to Ray Gatchell, Wilson (NC) Chamber of Commerce, November 11, 1943. Box 134.

91. Letter from Carl P. Russell, NPS Chief Naturalist, to Mrs. Frank Bierman, Bilssfield (MI), January 7, 1944. Box 134.

92. Letter from Isabelle F. Story, NPS Chief of Information, to *Pullman News*, Chicago, May 25, 1943. Box 134.

93. Letter from C. C. Mullady, Chief, NPS Contracts Division, to Audit Division, GAO, Washington (DC), November 24, 1943. Box 134.

94. Letter from Macnamee to Perry Edwards, NPS Finance Officer, March 1, 1943. Box 134.

95. Letter from Macnamee to Perry Edwards, NPS Acting Finance Officer, March 11, 1943. The letter was on the stationery of the DC office of Pan American Airways Systems. Box 134. After the war, Macnamee headed the publicity for the National Federation of American Shipping, later renamed the American Merchant Marine Institute. He and his wife were involved in a sensational society divorce trial in Washington in 1948. She accused him of drinking too much, and he accused her of laziness. A judge ruled that neither had legal grounds to justify divorce. "Court Holds Quarrels Trivial, Denies Divorce to Mrs. Macnamee," *WS*, October 14, 1948, B-4. Macnamee retired in 1957. "W.B. Macnamee, Maritime Institute Aide, to be Feted Tonight as Retirement Nears," *NYT*, July 30, 1957, 50. He died a year later. "Publicist MacNamee [sic] Dead at 68," *WP*, June 5, 1958, B2; "Bruce Macnamee, Publicist, is Dead," *NYT*, June 6, 1958, 23.

96. Memo from Macnamee to NPS Director, October 22, 1942. Box 135.

97. Morris, "Courtesy Pays Dividends," *CSM*, March 9, 1945, 18.

Notes to Chapter 8

1. Letter from Hillory A. Tolson, NPS Acting Director, to José Cidre, President, Corporacion Nacional Del Tourismo, Habana, Cuba, December 18, 1943. Box 134.

2. Letter from Macnamee to Perry Edwards, NPS Acting Finance Officer, March 11, 1943. Box 134. This was the second time he made that point. As part of his proposal in October 1942 to close USTB, he had explicitly told NPS's director that if the bureau resumed after the war, "I should like to direct this work again." Memo from Macnamee to NPS Director, October 22, 1942. Box 135.

3. Rice, "In the Field of Travel" (column), *NYT*, September 23, 1945, X-7.

4. Henry A. Wallace, "Tourism Important Factor in Economy of the World," *Boston Globe*, April 14, 1946, C3. Oddly, Wallace had been Truman's predecessor as vice president. When FDR ran for the Democratic nomination in 1932, one of his opponents was conservative House Speaker John Nance Garner (D-TX). After winning the nomination and in an effort to balance the ticket by appealing to conservatives, FDR invited Garner to be his running mate. Garner was never comfortable with the liberal tilt of the New Deal. During FDR's second term, the vice president broke with Roosevelt, particularly over the court-packing scheme. Assuming that FDR would not violate the two-term tradition for presidents, the estranged vice president began running for the 1940 Democratic nomination. When Roosevelt announced he would accept the nomination for a third term, Garner stayed in the race, forcing a vote by the Democratic convention. Thus, politically, it was impossible that he would continue as vice president. Wallace had been USDA secretary in FDR's first two terms, and Roosevelt picked him to be his running mate in 1940 because they were both liberal. However, Wallace was unpopular with many party bosses. When FDR was preparing to run for a fourth term in 1944, he felt he could not ignore the opposition to Wallace remaining as vice president. Eventually, Roosevelt dumped Wallace for the more conservative, but widely liked, Senator Harry S. Truman (D-MO). When FDR gave Wallace the bad news, he told Wallace that he could have any cabinet secretaryship he wanted for the fourth term. Wallace picked Commerce.

A few months after writing this column, in September 1946, Truman fired Wallace for making public comments about friendship with the USSR that were out of line with Truman's increasingly oppositional stance toward the Soviet Union, later called the Cold War. As a result, Wallace ran for president in 1948 as a third-party candidate with a more liberal platform than Truman's. Despite draining votes that would otherwise have gone to a Democrat, and to nearly everyone's surprise, Truman won.

5. AP, "Republican Economy Drive Scores Victory over Interior Budget: House Votes 45 Per Cent [*sic*] Funds Slash," *HC*, April 26, 1947, 1.

6. 61 *Stat.* 484.

7. Letter from Krug to Admiral W. W. Smith, Chairman, US Maritime Commission, August 29, 1947. Box 133.

8. For a complete list of federal and industry committee members, see "United States Travel Division Advisory Board," *Travel USA Bulletin* 1:1 (October 1948) 10–11.

9. Helen Ynonne Hughes, "Social and Personal" (column), *[Fayetteville] Northwest Arkansas Times*, September 10, 1947, 2.

10. National Park Service, *Directory–January 1947* (mimeograph), 34.

11. Jay Walz, "U.S. Travel Program," *NYT*, January 4, 1948, X-13; "All-Year Club Chief Named to U.S. Post," *LAT*, January 24, 1948, 2; "Former Altoonan Given Dismissal from State Job," *Altoona [PA] Mirror*, April 1, 1955, 29.

12. "U.S. Travel Division Collaborators," *Travel USA* 1:7 (April 1949) 10.

13. "Travel Agency Again Set Up," *[Phoenix] Arizona Republic*, October 19, 1947, 5; "Personals," *[Fayetteville] Northwest Arkansas Times*, October 21, 1947, 2. In 1963, when Interior and Commerce were still fighting over domestic travel promotion, a curt memo from NPS to Commerce's US Travel Service attached a copy of the 1947 press release as proof of Interior's extant jurisdiction. Memo from George B. Hartzog, NPS Associate Director, to John Black, Deputy Director, US Travel Service, May 15, 1963, Subject: "See America Now." File: 1319, Box 108, Hartzog Papers.

14. October 26, 1947: Rice, "The Field of Travel: U.S. Government Travel Bureau Revived" (column), *NYT*, X-15; Samuel Wallace, "U.S. Travel Division Gets a New Start," *CT*, G13. Also, "Bossemeyer Heads Department of Interior Travel Division," *Railway Age* 123:17 (October 25, 1947) 68.

15. Wallace, "Great Travel Industry Here for Convention," *CT*, October 22, 1947, 20; "Top Travel Officials to Meet in Chicago," *Boston Globe*, October 19, 1947, A-22.

16. "Travel Leaders Told U.S. Park Service's Plans," *CT*, October 24, 1947, 22.

17. "U.S. Travel Agency Restored to Assist Foreign Tourists," *CSM*, June 18, 1948, 11; "U.S. Travel Division Policy Is Outlined," *Boston Globe*, June 13, 1948, B-13; Beach Conger, "Travel Topics" (weekly column), *NYHT*, October 17, 1948, C13; "Opportunity," *Travel USA* 1:4 (January 1949) 2.

18. "Why Not Take a Fall Vacation?" *Changing Times: The Kiplinger Magazine* 3:10 (October 1949) 23.

19. AP, "Looking Ahead," *Bridgeport [CT] Post*, January 11, 1948, A-10.

20. Demonstrating USTD's symbiotic relationship with the industry, the National Association of Travel Organizations (NATO) had begun publishing a monthly magazine called *Travel USA* in 1946. The association ceased publication of its magazine when USTD started its own. In that respect, the governmental publication was merely a continuation of a private-sector one with their nearly identical names indicating this. This was a vivid example of the hand-in-glove relationship between private businesses and "their" federal agency.

21. Bossemeyer's predecessor, Macnamee, wrote an article for USTD's monthly: "Let's be Realistic," *Travel USA* 1:5 (February 1949) 8–9. He was identified as the executive assistant to the president, National Federation of American Shipping.

22. Based on entries in WorldCat/OCLC. Series: 269221755; no. 1: 76818747; no. 2: 76818756; no. 3: 855116418; no. 4: 84318490.

23. American Geographical Society, University of Wisconsin-Milwaukee, accessed January 7, 2020, http://collections.lib.uwm.edu/digital/collection/agdm/id/11451/rec/1. The map was such a hit that it continued to be mentioned in travel columns years after USTD had closed. Carol Lane, "Tips on Touring" (column), *Cleveland Plain Dealer*, February 27, 1953, 2; Charlotte Jenkins, "The Jenkins' 'Do It Yourself' Vacations," *Hyde Park [Chicago] Herald*, August 4, 1954, 6.

24. "World News Front," *Oakland [CA] Tribune*, April 7, 1948, C3.

25. "Session Discusses New Travel Ideas," *NYT*, October 24, 1948, S11. In 1949, Bossemeyer also spoke to the organization's mid-Atlantic chapter. Orval Hopkins, "Travel Time, and Open Road Calls," *WP*, April 17, 1949, L5. For a photo of Bossemeyer at another meeting with association officials, see Hopkins, "Getting a Lift in Portugal," *WP*, March 20, 1949, L6.

26. "N.A.T.O. Seeks Ways to Satisfy Travelers," *Boston Globe*, November 7, 1948, A-21. It will be recalled that one of Bossemeyer's first actions upon the reopening of USTB/USTD was to speak to the organization's annual conference in Chicago in October 1947.

27. "Northern Lakes Council Meets," *Escanaba [MI] Daily Press*, September 19, 1949, 7.

28. "Move Launched to Speed Data on River Parkway Road," *Alton [IL] Evening Telegraph*, February 3, 1950, 1.

29. "Dedication of Park at Signal Point on Walden's Ridge" (photo), online digital collection, Chattanooga (TN) Public Library. Lattimore filled in for Bossemeyer who had originally committed to attend. AP, "Signal Point Park Ceremonies Slated," *Kingsport [TN] Times*, April 28, 1948, 8. The *Herald Tribune* misreported that Bossemeyer had attended. "Chattanooga Area Adds New Scenic Attractions," *NYHT*, May 9, 1948, F22.

30. "Travel Art Exhibition," *Travel USA Bulletin* 1:1 (October 1948) 13; 1:2 (November 1948) 13.

31. "U.S. Official Eyes King Resort Study," *Asbury Park [NJ] Press*, February 25, 1949, 12.

32. "How to Make Your Vacation," *This Week* (April 24, 1949), *LAT*, F8; *NYHT*, SM-14.

33. 62 *Stat.* 1141.

34. "U.S. Travel Division Policy Is Outlined," *Boston Globe*, June 13, 1948, B-13. Also "U.S. Travel Agency Restored to Assist Foreign Tourists," *CSM*, June 18, 1948, 11.

35. Hopkins, "Get a Free Map, with Pictures," *WP*, May 30, 1948, L6.

36. W. Murray Metten, "National Association of Travel Officials," *Travel USA* 1:6 (March 1949) 4.

37. Wallace, "Press Drive to End Tax on Transportation," *CT*, January 30, 1949, E16.

38. Robert D. Byrnes, "Orders Still Unfilled on Plane Parts," *HC*, May 22, 1948, 2. He also argued that the subject was "academic" because transportation companies were already running at capacity, and they would not be able to accommodate any increased demand. "Washington Briefs: Travel Abroad Debated," *NYHT*, May 22, 1948, 9.

39. *CR* 94:7 (June 16, 1948) 8486. Oddly, the title of that section of the floor proceedings was "United States Travel *Service*" (emphasis added). Despite the

absence of major controversy or debate at that point in the legislative process, AP considered it important enough news to move a short story about it on its national wire. "U.S. Travel Bureau Bill Voted," *NYHT*, June 17, 1948, 28. The *Wall Street Journal* also reported on it. "Washington Checklist," June 17, 1948, 5.

40. The reason for the committee's inaction is undocumented. The committee's bill file (at the National Archives' Center for Legislative Archives) contained no helpful information or correspondence.

41. 80 H.R. 4204, *CR* 95:4 (April 12, 1949) 4445; 80 H.R. 4321, *CR* 95:4 (April 25, 1949) 5012. For the fate of the bills, see index for all House bills in that session, *CR* 95:17, 894, 897.

42. *CR* 95:3 (March 30, 1949) 3524–25. Attendance on the floor during debates on amendments was routinely sparse.

43. *CR* 95:9 (August 25, 1949) 12232–34.

44. Robert C. Albright, "Bills Involving $21 Billions [*sic*] Deadlocked," *WP*, September 12, 1949, 1, 6; United Press, "Interior Bill Passed," *NYT*, October 8, 1949, 28.

45. The death of USTD is particularly odd when contrasted with the Canadian government travel promotion in the same period. The national government there expanded considerably the work and funding of its travel bureau after WWII. I am indebted to one of the anonymous reviewers for noting this contrast and suggesting a possible explanation for it. The reviewer speculated that this may have reflected structural differences between the two countries, even though they are both federated systems. In the United States, the power of state travel offices and local chambers of commerce may have had more political weight and been more effective with their targeted and specific goals than any relatively feeble centralized effort by the US federal government. On the other hand, the Canadian federal government was more powerful and had more ability to act centrally and in a unified way compared to local efforts by individual provinces or cities.

46. The official was Robert Wall, deputy head of Commerce's International Trade Development Division in the Bureau of Foreign Commerce. Due to departmental rivalries, one could argue that he would have a motive in saying he was unaware of any such document because he would not want anything positive to be known about USTB/USTD's performance and legacy. Saying that could undercut Commerce's claim for the portfolio. However, he likely would have volunteered the existence of such a historical accounting if it reflected badly on USTB/USTD's record.

47. "Purely Personal," *CT*, November 20, 1949, F9; "U.S. Travel Division Discontinued," *Railway Age* 127:21 (November 19, 1949) 214. He briefly stayed on at Interior as NPS's chief of recreational planning. "New Officer Named by Travel Group," *Detroit Free Press*, March 26, 1950, D-17. Within a few months, he was hired by the National Association of Travel Agents to be its executive vice president. "Purely Personal," *CT*, March 12, 1950, G9. Bossemeyer worked there until his retirement in 1964. "People and Places," *CT*, February 16, 1964, H24. After

retiring, he moved back to California (where he had lived when heading USTB's San Francisco office in the early 1940s) and died there in 1969.

48. Wallace, "Canada Cancels Tax on Travel; Other Items," *CT*, April 17, 1949, G6.

49. "Purely Personal," *CT*, November 20, 1949, F9.

50. Hopkins, "Kissproof Kerchief Has Its Advantages" (Whistle Stopping weekly column), *WP*, November 13, 1949, L6.

Notes to Chapter 9

1. Jean Van Vranken, International News Service, "Trips to Marshall Plan Countries Urged by ECA," *Gladewater [TX] Daily Mirror*, July 27, 1949, 5; Frank Cipriani, "Rap Use of ECA Cash to Lure Tourists: U.S. Operators Hit Spending of Tax Money," *CT*, September 10, 1950, G5, G7. When considering Truman's legislative proposal for the Marshall Plan, conservatives in Congress insisted it be administered by a new and independent agency, not by the State Department (Steil 2018).

2. "Commerce Department Will Seek Funds for New World Travel Unit," *Wall Street Journal*, December 10, 1953, 13; Arthur Edson, AP, "Travel Urged to Replace American Aid with Trade," *HC*, January 21, 1954, 14A.

3. A 1956 article asserted that the office was "at present a one-man show" with Commerce trying to increase staffing to "a four-man operation." Nona Brown, "Tourism Program Drafted for U.S.," *NYT*, February 12, 1956, X-21.

4. "President Cited for Aid to Travel," *NYT*, May 11, 1954, 10. On another occasion, an assistant secretary of commerce said the department sought to "promote and encourage travel." Barnett D. Laschever, "Travel Agents Lauded for Good Will by Javits," *NYHT*, November 4, 1958, 17.

5. Laschever, "Need Seen for a U.S. Tourist Bureau," *NYHT*, June 1, 1958, D6, D10; Alec Henderson, "Program to Increase Travel Grows," *WP*, November 8, 1959, C16.

6. Paul J. C. Friedlander, "Plugging the Dollar Gap," *NYT*, March 13, 1960, XX-1, XX-15.

7. Laschever, "Need Seen for a U.S. Tourist Bureau," *NYHT*, June 1, 1958, D6.

8. United Press International, "U.S. Tourist Bill Voted," *NYT*, June 8, 1960, 27.

9. 75 *Stat.* 129–30.

10. Robert Berkvist, "Welcoming Committee," *NYT*, July 9, 1961, XX-1; Horace Sutton, "U.S. Will Enter the Rubberneck Business," *NYHT*, September 10, 1961, C18.

11. Discounting for inflation, USTS's 1979 budget was about double its 1961 funding level.

12. 78 *Stat.* 388.

13. Letter from Hartzog to Ralph A. Dungan, Special Assistant to the President, September 14, 1964. File 1319, Box 108, Hartzog Papers.

14. Memo from Hartzog to Udall, February 4, 1965, Subject: "See the U.S.A." Ibid.

15. *CR* 111:8 (May 12, 1965) 10401–02.

16. Text of speech, May 25, 1967. Speeches File, Hubert H. Humphrey Papers, Manuscript Collections, Minnesota Historical Society, accessed January 7, 2020, http://www2.mnhs.org/library/findaids/00442/pdfa/00442-02237.pdf.

17. "Industry-Government Special Travel Task Force," November 16, 1967, *Weekly Compilation of Presidential Documents* 3:46 (November 20, 1967) 1581. The original mimeographed press release is in the papers of Shooshan's boss, Under Secretary David S. Black. File 13, Box 72, David S. Black Papers, Kennedy Presidential Library.

18. Given that everything in Washington is political, this was occurring when Johnson was running for reelection and during the Tet Offensive in Viet Nam. The New Hampshire primary was about a month away, scheduled for March 12, 1968.

19. President Johnson proposed to Congress to create DOT by consolidating dispersed federal units. Congress approved the bill in 1966, and DOT came into existence in April 1967.

20. 90 H.R. 10393, *CR* 113:11 (May 25, 1967) 14229. Hartzog interpreted the failure of the bill as proving the great difficulty in changing the status quo with new legislation on the subject.

21. The most plausible explanation (Clark 2013, 48) is that, on behalf of Secretary Udall and Hartzog, Shooshan pushed hard for this to be included in the report's recommendations and that it was added very late in the drafting process. Regrettably, the Johnson Presidential Library had no records of the internal deliberations of the task force nor early drafts of its final report. Emails to the author from Jenna De Graffenried, Archivist, Johnson Presidential Library, March–April, 2018, author's files.

22. Letters from George E. Robinson, Deputy Assistant Secretary of the Interior to Representative Julia Butler Hansen, Chairman, Subcommittee on Interior, House Appropriations Committee and Senator Carl Hayden, Chairman, Senate Appropriations Committee, April 9, 1969. File 1319, Box 108.

23. Responses from Butler, April 18, 1968; Hayden, April 22, 1968. Ibid.

24. He had barely squeaked through a win over McCarthy in the New Hampshire primary on March 12, 49 percent to 42 percent. The Wisconsin primary, scheduled for April 2, was looming and not looking good. LBJ's political standing was so bleak that his campaign office in Milwaukee, for lack of the usual flow of volunteers, had to *hire* temps to staff it. A Milwaukee native, I was then an undergrad at the University of Wisconsin-Madison. Along with many other students, I was a volunteer in the McCarthy campaign in the winter of 1967–68. Based on

the motto "Get Clean for Gene," we got haircuts to shorten what at the time was considered long hair and we also trimmed or shaved off our facial hair. I then went door-to-door talking to voters in the rural counties surrounding Madison. It was cold!

25. George Romney, Michigan governor and father of Mitt Romney, the Republican nominee for president in 2012 and, later, US senator, had already dropped out.

26. Memo from Udall to bureau chiefs, Subject: Interior Task Force on Travel, July 23, 1968. File 1319, Box 108.

27. "A Report to the Secretary of Interior; From the Interior Task Force on Travel," First Draft, September 8, 1969, p. 7a. Ibid.

28. Memo from Hartzog to Shooshan, Subject: Draft of Report of Interior Task Force on Travel, September 20, 1968. Ibid.

29. Memo from William C. Everhart, NPS Assistant Director for Interpretation, to Shooshan, Subject: Draft of Task Force Report, September 10, 1968. Ibid.

30. "A Report to the Secretary of the Interior from the Interior Task Force on Travel," n.d., p. 5. The cover memo was dated September 30, 1968. Ibid.

31. Indicating how rushed this effort was to take advantage of the final months of Johnson's presidency, Udall apparently was out of town when the report was formally submitted to him. No one wanted to lose any time waiting until he got back. The report was signed and approved on October 4, 1968, by the acting secretary, Under Secretary Black. Ibid.

32. 90 S. 4134, *CR* 114:22 (October 3, 1968) 29279. The identical bill in the House, 90 H.R. 20505 was introduced by Representative John Saylor (R-PA), *CR* 114:24 (October 11, 1968) 30817. Significantly, the Senate bill was referred to the Interior Committee, a group predisposed to support NPS. However, the House bill was sent to the Commerce Committee, which was oriented to favor the Commerce Department and its travel portfolio.

33. Department of Interior press releases, October 22, 1968: "Udall Announces Creation of Office of Tourist Development to Promote More Travel" and "Excerpts from Remarks of Secretary." File 1319, Box 108.

34. "Butterfield Named Director of Travel for National Park Service" (press release), October 24, 1968. Ibid. The travel sections of two newspapers treated it as news: "Butterfield to New Post," *WS*, November 3, 1968, B-6; "People and Places," *CT*, December 1, 1968, M-12.

35. Secretary's Order No. 2912, November 4, 1968, prepared for publication in the *Federal Register*. File 1319, Box 108.

36. *Official Congressional Directory*, 91st Congress, 1st Sess. (January 1969), p. 511.

37. Letter from Hickel to Martin F. Schafer, Director, Division of Travel, Alaska Department of Economic Development and Planning, April 14, 1969. File 1320, Box 108. United Press International, "Hickel Office Urges Alaska Vacations," *NYT*, February 16, 1969, 29.

38. Letters signed by Udall, November 8 and 22, 1968. File 1319, Box 108.

39. "Udall Names Travel Advisory Board" (press release), December 14, 1968. File: Travel Advisory Board, Folder 20, Box 14, Subseries 1b, Series 1, Hickel Papers.

40. "Udall Appoints Washingtonians," *WS*, December 29, 1968, C-15; "Travel Board," *Baltimore Sun*, January 5, 1969, H6.

41. I did not locate the original letter. The new Republican Secretary of Commerce, Maurice Stans, referred to it in a letter to the new Interior Secretary, Walter Hickel on February 11, 1969. File 1320, Box 108.

42. The letter falsely claimed that USTD had been defunded due to the Korean War. But, as noted in chap. 8, USTD closed in November 1949, while the Korean War did not start until mid-1950.

43. Memo from Hartzog to Hickel, Subject: Travel Advisory Board Meeting, February 4th, January 28, 1969. File: Travel Advisory Board, Folder 20, Box 14, Subseries 1b, Series 1, Hickel Papers. It is unclear from archival records if Hickel attended.

44. February 16, 1969: "Hickel's Office Boosts Alaska," *Boston Globe*, A-6; UPI, "Hickel Office Urges Alaska Vacations," *NYT*, 29.

45. Letter from Secretary of Commerce Maurice Stans, to Hickel, February 11, 1969. File 1320, Box 108.

46. As part of his testimony, Hartzog submitted to the subcommittee a document titled "Travel Program Activities—1940 Travel Act." According to the hearing record, the document (which was not included in the published hearing), "may be found in the committee files" (US House 1970a, 111). However, the document could not be located at the Center for Legislative Archives. Email to the author from Adam Berenbak, Archivist, July 12, 2018. Author's files.

47. The speech was to an environmental conference at the University of Wisconsin-Madison in the summer of 1969. Butterfield was a Wisconsin alum.

48. Eric Friedheim, "Travel Outlook in Washington" (opinion), *Travel Agent* 75:1 (November 25, 1968) 70. Friedheim was the publication's editor.

49. Letter from Hickel to Staggers, August 12, 1969. File 1320, Box 108.

50. Memo from Butterfield to Hartzog, Subject: Letter of February 11 from Secretary of Commerce to Secretary of the Interior, March 5, 1969; Memo from Hartzog to Butterfield et al., Subject: Meeting with Mr. Langhorne Washburn [USTS Director], June 26, 1969; NPS draft of unsigned Memorandum of Agreement Between USTS and NPS, prepared for Hartzog, n.d. (probably late summer or early fall, 1969). File 1320, Box 108.

51. *CR* 116:29 (December 1, 1970) 39241–42.

52. 84 *Stat.* 1437.

53. Ten days later (and after Hickel's letter to Nixon), there was another police shooting at Jackson State University in Mississippi, a predominantly African American institution. Allegedly responding to projectiles thrown at them, officers of the State Highway Patrol shot at the windows of a dormitory and killed two students.

54. "Text of the Hickel Letter," *NYT*, May 7, 1970, 18.

55. As a summer intern at the US Geological Survey in 1970, I was in the audience when he spoke in the Department's auditorium to a very large crowd of Interior's civil servants. (It was not a public event and was closed to the media.) He declared that he would not be intimidated by administration hints to tone down his public criticisms and to keep them private. He was wildly applauded by the civil servants who craved, above all, autonomy from what they considered political interference.

56. It is possible that the administration felt that few Americans would be paying attention to political news during the long holiday weekend, and therefore the timing would minimize any potential political backlash.

57. In Nixon's second term, Malek was appointed deputy director of OMB. He spearheaded the effort to impose a technique then-popular in the business sector, Management-by-Objectives (MBO), onto the federal executive branch (Lee 2010, 99–100).

58. *NYT*: "Two Hickel Policies Reported Reversed," November 28, 1970, 24; "Interior Official Balks at Quitting," November 30, 1970, 21. Glasgow's position was vacant until the next spring, when he was succeeded by Nathaniel Reed. In his memoir, Reed said that Glasgow's firing had been "senseless" (2016, 211).

59. Ken W. Clawson, "No More Firings Planned at Interior Dept.," *WP*, December 1, 1970, A2. A few months later, Clawson resigned from the *Post* to become a White House aide to President Nixon in the Office of Communications.

60. Confusingly, the term human resources changed meaning over time. In the 1970s, it was a newish synonym for federal programs in the areas of health, education, and welfare. In the twenty-first century, it came to mean personnel management. This latter meaning was so universally accepted that the acronym "HR" was often used without further explanation.

61. The administration's bill to create a Department of Community Development almost passed. It was approved by a House committee and recommended for passage. But in the summer of 1972, the House Democratic leadership decided not to send the bill to the floor for a vote in order to deny Nixon a political and legislative win to boast about during his reelection campaign.

62. William M. Blair, "Morton to Present Plan to Reorganize Interior Department," *NYT*, March 11, 1971, 17.

63. The date of Morton's letter in the congressional hearing record appears to be a typo. The date on an archival copy of the letter is April 15, 1971. File 1321, Box 108. By the time the new assistant secretary overseeing NPS, Nathaniel Reed, took office in May 1971, Morton had already made the authoritative decision that NPS should give up its travel portfolio to Commerce. Reed makes no mention of NPS's travel programming in his memoir (2016).

64. The premise of a congressional veto reversed the normal pattern of American government because in this case the executive was enacting something

akin to legislation, and the legislative branch had an executive-type power of only being able to agree to it or veto it. The concept of a legislative veto was eventually ruled unconstitutional by the Supreme Court in 1983 in its *INS v. Chadha* decision.

65. 83 *Stat.* 6.

66. Nixon submitted the Action reorganization plan before April 1, but floor votes on the disapproval motions did not occur until later in the year, after April 1. These later actions were valid because the expiration of the law related to the deadline for *submitting* plans, not acting on them.

67. 85 *Stat.* 576.

68. It is also possible that later in-house legal reviews of Morton's draft could have raised the question whether the transfer would violate the limitations on reorganization plans by tampering with the original statutory authority to Interior to operate USTB. If so, then a president's reorganization powers would be precluded from making such a change.

69. Letter from Butterfield to C. Langhorne Washburn, Assistant Secretary of Commerce for Tourism, September 28, 1971. File 1321, Box 108.

70. Under the plan, Nixon appointed USDA Secretary Earl Butz to a second role as the super-secretary for natural resources. His domain included all of Interior. This meant that Morton was theoretically subordinate to Butz. According to Nathan, Morton was hurt and disappointed not to get the appointment but kept his feelings to himself and stayed on (1986, 52n20). Nixon's administrative reorganization also included the expectation of appointing George Schultz, treasury secretary, to a second post as assistant to the president for economic affairs. This would be somewhat akin to being a supersecretary for a department of economic affairs. Commerce, including USTS, would have been under his jurisdiction. However, Schultz did not want to exercise any administrative power over such a conglomerate. Rather, he was only interested in coordinating *policy*, particularly international economic policy (Lee 2012, 20–23, 206–07). Therefore, the new commerce secretary for the second term, Frederick Dent, did not report to Schultz in the way that Morton was expected to report to Butz. In May 1973, as a result of Watergate and the resignation of John Ehrlichman, the top White House domestic assistant, Nixon killed the administrative reorganization. In remaking Nixon's second-term cabinet into his own, in 1975, President Ford moved Morton to Commerce. This meant Morton became overseer of USTS, including the domestic travel responsibilities it now had due to Morton's support for that transfer when he was secretary of interior.

71. December 7, 1972: Peter Braestrup, "Park Service Chief Fired in Shakeup," *WP*, A1, A14; William M. Blair, "8 are Dismissed from High Posts in Interior Dept.," *NYT*, 1, 62. Significantly, of the eight people Morton fired that day (mostly assistant secretaries), Hartzog was the only one who was not a political appointee. The only assistant secretary he retained was Nathaniel Reed, who came into office in May 1971 as a Morton appointee (Reed, 2016, 272). Reed and Morton felt that Hartzog had undermined them too many times. In one case, Reed complained that

Hartzog had lobbied congressional leaders "without following rigorous departmental procedures," that is, had done so behind their backs (2016, 92–93). Reed did not like that Hartzog "had a close working relationship with many of the Democratic senators" (192), which he saw as likely disloyalty to administration policies. In his memoir, Reed said, "my relationship with Director Hartzog disintegrated over his continued efforts to undermine the secretary" (95).

72. In the hindsight of history, the most famous of these appointments was of Alexander Butterfield's (no relation to Ben Butterfield). A White House aide in the first term, Nixon named him to head the Federal Aviation Administration (FAA). Butterfield was the person who later revealed to the Senate Watergate Committee the existence of a taping system in the Oval Office.

73. Linda Charlton, "Keogh, Former Aide to Nixon, Is Chosen as Head of U.S.I.A.," *NYT*, December 14, 1972, 1.

74. Walker had briefly served as an assistant to Secretary Hickel in 1969. However, he was untainted by that association because he had moved to the White House in January 1970, before Hickel had broken with Nixon (US House 1973b, 201).

75. Morton asked Walker to resign a month after Nixon had resigned. "Parks Director Quits after Senate Inquiry," *LAT*, September 12, 1974, A4.

76. 93 H.R. 4964, *CR* 119:5 (February 28, 1973) 5982.

77. *CR* 119:30 (December 3, 1973) 39246.

78. Inouye led the Senate's consideration of the bill that included the transfer (US Senate 1973).

79. "Meanwhile, Back in the United States," *NYT*, April 22, 1973, XX-7, emphasis added. The nonbylined column was a sidebar to an article about how Canada had a vigorous governmental program to encourage its youth to explore their country. Steven Kinzer, "Youth in Canada: Drifting Around with the Government's Blessing," XX-1, 7.

80. 87 *Stat.* 765.

81. Kathy Sawyer, "On Capitol Hill" (column), *WP*, October 2, 1981, A7.

82. 95 *Stat.* 1011–18.

83. Alfred Borcover, "Tourism Funding Cut by Reagan," *HC*, May 12, 1985, F4.

84. Robert Pear, "Most in U.S. Will Feel Effect of Shift in Spending Priorities," *NYT*, October 28, 1990, 26.

85. 124 *Stat.* 56–65. Even though he signed it in 2010, the law's official name was the Travel Promotion Act of 2009.

86. P.L. 116-94, pp. 495–96.

87. "NTTO Staff," accessed January 7, 2020, http://tinet.ita.doc.gov/about/overview.asp.

88. US Office of Management and Budget, "Budget of the US Government, Fiscal Year 2021: Appendix," February 2020, p. 1343.

Notes to Conclusion

1. Section 3, 54 *Stat.* 773.
2. "Uncle Sam, Travel Guide" (editorial), *Springfield [MA] Union*, December 14, 1937. Box OV 27, Ickes Papers.
3. "Another Bureau" (editorial), *Manchester [NH] Union*, June 3, 1939. Box OV 40, Ickes Papers.
4. Robert Pear, "Most in U.S. Will Feel Effect of Shift in Priorities," *NYT*, October 28, 1990, 26.
5. The quip is usually attributed to Senator Russell B. Long (D-LA) in 1973. However, slightly different variations had existed before that.

Bibliography

Primary and Archival Sources

Black, David S., Personal Papers (#31). Kennedy Presidential Library, Boston, MA.
Hartzog, George B., Jr., Papers, Mss.0074. Special Collections and Archives, Library, Clemson University, Clemson, SC.
Hickel, Walter J., Papers. Archives and Special Collections, Consortium Library, University of Alaska-Anchorage.
Humphrey, Hubert H., Papers. Speech Text File, Minnesota Historical Society. Accessed January 7, 2020: http://www2.mnhs.org/library/findaids/00442.xml.
Ickes, Harold L. (1953) 1974. *The Secret Diary of Harold L. Ickes; Vol. II: The Inside Struggle, 1936–1939*. New York: Da Capo Press.
Ickes, Harold L., Papers of. Manuscript Division, Library of Congress, Washington, DC.
National Park Service, Central Files (Record Group [RG] 79). National Archives II, College Park, MD.
Publications of the Federal Government (RG 287). National Archives II.
Roosevelt, Franklin D., President's Secretary's File (PSF), FDR Library, Hyde Park, NY.
Roosevelt, Franklin D. (1938–1950) 1969. *The Public Papers and Addresses of Franklin D. Roosevelt*, 13 vols. New York: Russell and Russell.
Roosevelt, Franklin D. 1972. *Complete Presidential Press Conferences of Franklin D. Roosevelt*. New York: Da Capo Press.
Smith, Harold D., Papers of. FDR Library, Hyde Park, NY.
US House of Representatives, Records of (RG 233). Center for Legislative Archives, National Archives, Washington, DC.
US Senate, Records of (RG 46). Center for Legislative Archives, National Archives, Washington, DC.

Federal Publications

Bratter, Herbert M. 1931. *The Promotion of Tourist Travel by Foreign Countries*. Trade Promotion Series No. 113, US Department of Commerce, December 1930. Washington, DC: Government Printing Office (GPO).

Carter, Jimmy. 1982. *Public Papers of the Presidents of the United States: Jimmy Carter, 1980–81; Book III*. Washington, DC: GPO.

Dorsett, Harold L. 1938. *Pleasure Travel and Its Relationship to the Automobile Industry*. Washington, DC: US Travel Bureau.

———. 1940. *Inter-American Travel*. Washington, DC: US Travel Bureau.

———, and Eugene I. Johnson. 1940. *Recreational Travel and Land Use*. Washington, DC: US Travel Bureau.

Ickes, Harold L. 1937. *Annual Report of the Secretary of the Interior for the Fiscal Year Ending June 30, 1937*. Washington, DC: GPO.

———. 1941. *Annual Report of the Secretary of the Interior; Fiscal Year Ended June 30, 1941*. Washington, DC: GPO.

———. 1942. *Annual Report of the Secretary of the Interior; Fiscal Year Ended June 30, 1942*. Washington, DC: GPO.

Industry-Government Special Task Force on Travel. 1968. *Report to the President of the United States*. Washington, DC: GPO.

Johnson, Lyndon B. 1970. *Public Papers of the Presidents of the United States: Lyndon B. Johnson, 1968–69; Book I*. Washington, DC: GPO.

Kennedy, John F. 1962. *Public Papers of the Presidents of the United States: John F. Kennedy, 1961*. Washington, DC: GPO.

US Bureau of the Budget. 1942, January 21. *Organization Nomenclature in the Federal Government*, mimeograph. Washington, DC: BOB.

US Commission on Organization of the Executive Branch of the Government (aka the First Hoover Commission). 1949, January. Task Force on Natural Resources. *Organization and Policy in the Field of Natural Resources: A Report with Recommendations, Appendix L*. Washington, DC: GPO.

US Congress. 1942. Joint Committee on Reduction of Nonessential Federal Expenditures. *Reduction of Nonessential Federal Expenditures*, Part 4, hearing. 77th Cong., 2nd sess.

US Department of Commerce. 1968. *Travel USA*. Series on "Do You Know Your Economic ABCs?," 9th issue. Washington, DC: GPO.

US House. 1931. Committee on Interstate and Foreign Commerce. *Bureau of Foreign and Domestic Commerce Travel Division*, hearing. 71st Congress, 3rd sess.

———. 1934. *Aid in Providing the People of the United States with Adequate Facilities for Park, Parkway, and Recreational-Area Purposes, etc.* 73rd Cong., 2nd sess., H. Rep. 1895.

———. 1935a. Committee on Interstate and Foreign Commerce, Subcommittee [n.t.]. *Tourist Travel Division*, hearing. 74th Cong., 1st sess.

———. 1935b. *To Encourage Travel to and within the United States*. 74th Congress, 1st sess., H. Rep. 1500.

———. 1936. *United States Travel Commission*. 74th Cong., 2nd sess., H. Rep. 2254.

———. 1938. Committee on Appropriations, Subcommittee on Interior Department. *Interior Department Appropriation Bill for 1939*, Part I, hearings. 75th Cong., 3rd sess.

———. 1939a. Committee on Appropriations, Subcommittee on Interior Department. *Interior Department Appropriation Bill for 1940*, Part I, hearings. 76th Cong., 1st sess.

———. 1939b. Committee on Appropriations, Subcommittee in Charge of Deficiency Appropriations. *First Deficiency Appropriation Bill for 1939*, hearings. 76th Cong., 1st sess.

———. 1939c. Committee on Interstate and Foreign Commerce, Subcommittee (n.t.). *National Travel Board*, hearing. 76th Cong., 1st sess.

———. 1939d. *Encourage Travel in the United States*. 76th Cong., 1st sess., H. Rep. 395.

———. 1939e. *Encouraging Travel in the United States*. 76th Cong., 1st sess., H. Rep. 1010.

———. 1940a. Committee on Appropriations, Subcommittee on Interior Department. *Interior Department Appropriation Bill for 1941*, Part I, hearings. 76th Cong., 3rd sess.

———. 1940b. Committee of Conference. *Travel Promotion Act*. 76th Cong., 3rd sess., H. Rep. 2764.

———. 1941a. Committee on Appropriations, Subcommittee on Interior Department. *Interior Department Appropriation Bill for 1942*, Part I, hearings. 77th Cong., 1st sess.

———. 1941b. *The Budget of the United States Government for the Fiscal Year Ending June 30, 1942*. 77th Cong., 1st sess., H. Doc. 28.

———. 1941c. Committee on Appropriations. *Interior Department Appropriation Bill, 1942*. 77th Cong., 1st sess., H. Rep. 476.

———. 1941d. Committee of Conference. *Department of the Interior Appropriation Bill, 1942*. 77th Cong., 1st sess., H. Rep. 773.

———. 1941e. Committee on Appropriations, Subcommittee on State, Justice, and Commerce Departments Appropriations. *Department of State Appropriation Bill for 1942*, hearings. 77th Cong., 1st sess.

———. 1941f. *State, Commerce, Justice, and the Judiciary Appropriation Bill, Fiscal Year 1942*. 77th Cong., 1st sess., H. Rep. 360.

———. 1942a. *The Budget of the United States Government for the Fiscal Year Ending June 30, 1943*. 77th Cong., 2nd sess., H. Doc. 528.

———. 1942b. Committee on Appropriations, Subcommittee on Interior Department. *Interior Department Appropriation Bill for 1943*, Part I, hearings. 77th Cong., 2nd sess.

———. 1942c. *Interior Department Appropriation Bill, 1943.* 77th Cong., 2nd sess., H. Rep. 1935.
———. 1942d. Committee of Conference. *Interior Department Appropriation Bill, 1943.* 77th Cong., 2nd sess., H. Rep. 2260.
———. 1942e. Committee of Conference. *Interior Department Appropriation Bill, 1943.* 77th Cong., 2nd sess., H. Rep. 2294.
———. 1943a. *The Budget of the United States Government for the Fiscal Year Ending June 30, 1944.* 78th Cong., 1st sess., H. Doc. 27.
———. 1943b. Committee on Appropriations, Subcommittee on Interior Department. *Interior Department Appropriation Bill for 1944*, Part I, hearings. 78th Cong., 1st sess.
———. 1944. Committee on Appropriations, Subcommittee on Interior Department. *Interior Department Appropriation Bill for 1945*, Part 1, hearings. 78th Cong., 2nd sess.
———. 1945. Committee on Appropriations, Subcommittee on Interior Department. *Interior Department Appropriation Bill for 1946*, Part 1, hearings. 79th Cong., 1st sess.
———. 1946. Committee of Conference. *Interior Department Appropriation Bill, 1947.* 79th Cong., 2nd sess., H. Rep. 2329.
———. 1947a. *The Budget of the United States Government for the Fiscal Year Ending June 30, 1948.* 80th Cong., 1st sess., H. Doc. 19.
———. 1947b. Committee on Appropriations, Subcommittee on Interior Department. *Interior Department Appropriation Bill for 1948*, Part 1, hearings. 80th Cong., 1st sess.
———. 1947c. *Interior Department Appropriation Bill, 1948.* 80th Cong., 1st sess., H. Rep. 279.
———. 1948a. *The Budget of the United States Government for the Fiscal Year Ending June 30, 1949.* 80th Cong., 2nd sess., H. Doc. 456.
———. 1948b. Committee on Appropriations, Subcommittee on Interior Department. *Interior Department Appropriation Bill for 1949*, Part 1, hearings. 80th Cong., 2nd sess.
———. 1948c. *Interior Department Appropriation Bill, 1949.* 80th Cong., 2nd sess., H. Rep. 2038.
———. 1948d. Committee of Conference. *Department of the Interior Appropriation Bill, 1949.* 80th Cong., 2nd sess., H. Rep. 2398.
———. 1948e. Committee on Interstate and Foreign Commerce. *Travel in the United States*, hearing. 80th Congress, 2nd sess.
———. 1948f. *United States Travel Bureau.* 80th Congress, 2nd sess., H. Rep. 2221.
———. 1949a. Committee on Appropriations, Subcommittee on Interior Department. *Interior Department Appropriation Bill for 1950*, Part 1, hearings. 81st Cong., 1st sess.

———. 1949b. *Interior Department Appropriation Bill, 1950.* 81st Cong., 1st sess., H. Rep. 324.

———. 1949c. Committee of Conference. *Interior Department Appropriation Bill, 1950.* 81st Cong., 1st sess., H. Rep. 1380.

———. 1954. Committee on Foreign Affairs, Subcommittee on Foreign Economic Policy. *To Promote the Foreign Policy of the United States by Fostering International Travel and the Exchange of Persons,* hearings. 83rd Cong., 2nd sess.

———. 1960. Committee on Interstate and Foreign Commerce, Subcommittee on Commerce and Finance. *Office of International Travel,* hearing. 86th Cong., 2nd sess.

———. 1968. Committee on Appropriations, Subcommittee on Department of the Interior and Related Agencies. *Department of the Interior and Related Agencies Appropriations for 1969,* Part 2, hearings. 90th Cong., 2nd sess.

———. 1969a. *The Budget of the United States Government, Fiscal Year 1970; Appendix.* 91st Cong., 1st sess., H. Doc. 91–16.

———. 1969b. Committee on Appropriations, Subcommittee on Department of the Interior and Related Agencies. *Department of the Interior and Related Agencies Appropriations for 1970,* Part 2, hearings. 91st Cong., 1st sess.

———. 1969c. Committee on Appropriations. *Department of the Interior and Related Agencies Appropriation Bill, 1970.* 91st Cong., 1st sess., H. Rep. 91–361.

———. 1969d. Committee of Conference. *Appropriations for [Interior Department for] Fiscal Year Ending June 30, 1970.* 91st Cong., 1st sess., H. Rep. 91–570.

———. 1970a. Committee on Interstate and Foreign Commence, Subcommittee on Commerce and Finance. *U.S.A. Travel Promotion,* hearings. 91st Cong., 1st and 2nd sess.

———. 1970b. *Domestic Travel Promotion.* 91st Cong., 2nd sess., H. Rep. 91–977.

———. 1970c. *International Travel Promotion.* 91st Cong., 2nd sess., H. Rep. 91–976.

———. 1971. *The Budget of the United States Government, Fiscal Year 1972; Appendix.* 92nd Cong., 1st sess., H. Doc. 92–16.

———. 1973a. Committee on Interstate and Foreign Commence, Subcommittee on Commerce and Finance. *International Travel Act Authorization and Domestic Travel Authority Transfer,* hearing. 93rd Cong., 1st sess.

———. 1973b. Committee on Appropriations, Subcommittee on Department of the Interior and Related Agencies. *Department of the Interior and Related Agencies Appropriations for 1974,* Part 2, hearings. 93rd Cong., 1st sess.

———. 1973c. *International Travel Act of 1961 Authorization.* 93rd Cong., 1st sess., H. Rep. 93–651.

———. 1980. Committee of Conference. *National Tourism Policy Act.* 96th Cong., 2nd sess., H. Rep. 96–1532.

———. 1981. *National Tourism Policy Act.* 97th Cong., 1st sess., H. Rep. 97–107, Part 1.

———. 1982. *International Travel Act Authorization for 1983*. 97th Cong., 2nd sess., H. Rep. 97–568.

US National Park Service. 1938. *1937 Yearbook: Park and Recreation Progress*. Washington, DC: GPO.

US National Tourism Resources Review Commission. 1973. *Destination USA: Report; Volume 4: Federal Role*. [Washington, DC]: [National Tourism Resources Review Commission].

US Office of Management and Budget, Executive Office of the President. 1971 (March). *Papers Relating to the President's Departmental Reorganization Program: A Reference Compilation*. Washington, DC: OMB. (In February 1972, OMB released a revised version of the president's plan. To avoid confusion between the two editions, the 1971 version had a grey cover and the 1972 cover was green.)

US Senate. 1935a. Committee on Commerce. *United States Travel Commission*, hearing. 74th Cong., 1st sess.

———. 1935b. *United States Travel Commission*. 74th Congress, 1st sess., S. Rep. 999.

———. 1938. *Travel Bill*. 75th Cong., 3rd sess., S. Rep. 1671.

———. 1939. Committee on Commerce, Subcommittee on General Reference. n.t. [To Encourage Travel in the U.S. and for Other Purposes], hearing. 76th Cong. 1st sess. Unpublished, ProQuest Congressional ID number: HRG-1939-COM-0017.

———. 1940a. *Supplemental Estimate of Appropriation, Department of State*. 76th Cong., 3rd sess., S. Doc. 244.

———. 1940b. *Encouraging Travel in the United States*. 76th Congress, 3rd sess., S. Rep. 1333.

———. 1941a. Committee on Appropriations, Subcommittee [n.t.]. *Interior Department Appropriation Bill for 1942*, hearings. 77th Cong., 1st sess.

———. 1941b. *Interior Department Appropriation Bill, 1942*. 77th Cong., 1st sess., S. Rep. 366.

———. 1941c. Committee on Appropriations, Subcommittee [n.t.]. *Departments of State, Commerce, Justice, and the Federal Judiciary Appropriation Bill for 1942*, hearing. 77th Cong., 1st sess.

———. 1942a. Committee on Appropriations, Subcommittee [n.t.]. *Interior Department Appropriation Bill for 1943*, hearings. 77th Cong., 2nd sess.

———. 1942b. *Interior Department Appropriation Bill, 1943*. 77th Cong., 2nd sess., S. Rep. 1380.

———. 1946a. Committee on Appropriations, Subcommittee [n.t.]. *Interior Department Appropriation Bill for 1947*, hearings. 79th Cong., 2nd sess.

———. 1946b. *Supplemental Estimates—Department of the Interior, Communication from the President of the United States*. 79th Cong., 2nd sess., S. Doc. 194.

———. 1946c. *Interior Department Appropriation Bill, 1947*. 79th Cong., 2nd sess., S. Rep. 1434.

———. 1948a. Committee on Appropriations, Subcommittee [n.t.]. *Interior Department Appropriation Bill for 1949*, hearings. 80th Cong., 2nd sess.

———. 1948b. *Interior Department Appropriation Bill, 1949.* 80th Cong., 2nd sess., S. Rep. 1609.

———. 1949. Committee on Appropriations, Subcommittee [n.t.]. *Interior Department Appropriation Bill for 1950*, Part 1, hearings. 81st Cong., 1st sess.

———. 1960a. Committee on Interstate and Foreign Commerce, Subcommittee on Foreign Commerce. *Foreign Commerce Study (International Travel)*, hearings. 86th Cong., 2nd sess.

———. 1960b. *International Travel.* 86th Cong., 2nd sess., S. Rep. 1493.

———. 1968. Committee on Appropriations, Subcommittee [n.t.]. *Department of the Interior and Related Agencies Appropriations for Fiscal Year 1969*, Part 1, hearings. 90th Cong., 2nd sess.

———. 1969a. Committee on Appropriations, Subcommittee [n.t.]. *Department of the Interior and Related Agencies Appropriations for Fiscal Year 1970*, hearings. 91st Cong., 1st sess.

———. 1969b. Committee on Appropriations. *Interior Department and Related Agencies Appropriations Bill, 1970.* 91st Cong., 1st sess., S. Rep. 91–420.

———. 1970. Committee on Commerce. *Domestic Travel Promotion.* 91st Cong., 2nd sess., S. Rep. 91–1373.

———. 1971. Committee on Appropriations, Subcommittee [n.t.]. *Department of the Interior and Related Agencies Appropriations for Fiscal Year 1972*, hearings. 92nd Cong., 1st sess.

———. 1973. *Authorization for the United States Travel Service.* 93rd Cong., 1st sess., S. Rep. 93–195.

US Tourist Bureau. 1937a. *The United States Tourist Bureau* (10 pages, full sheet size). Washington, DC: GPO.

———. 1937b. *The United States Tourist Bureau, Washington, D.C.* (9 pages, half sheet size). Washington, DC: Department of Interior.

US Travel Bureau. 1938a. *Directory of Travel Agencies in the United States, 1938.* Washington, DC: Branch of Recreational Planning and State Co-operation, National Park Service, US Department of the Interior.

———. 1938b. *Bibliography and List of Associations and Manufacturers of Equipment—Tourist Camps, Cabins, Auto Courts and Trailers*, mimeograph. Washington, DC: USTB, National Park Service, US Department of the Interior.

———. 1939a. *Production Procedure for a National Brochure*, typescript. Washington, DC: National Park Service, Department of Interior.

———. 1939b. *A Travel Service*, typescript. Washington, DC: Department of Interior.

———. 1939c. *Calendar of Events, 1939: Second Half, July Through December*, typescript. Washington, DC: National Park Service, Department of Interior.

———. 1939d. *State Travel Information Agencies.* Washington, DC: Department of Interior.

———. 1939e. *Directory of Negro Hotels and Guest Houses in the United States, 1939.* Washington, DC [?]: National Park Service.
———. 1940a. *A Travel Service*, typescript. Washington, DC: Department of Interior.
———. 1940b. *Preliminary Descriptive Poster List.* Washington, DC: USTB.
———. 1940c. *Winter Sports Round Up.* New York: USTB.
———. 1940d. *Ski West, 1940–41.* San Francisco: USTB.
———. 1941a. *Ski West, 1941–42.* San Francisco: USTB.
———, 1941b. *Calendar of Events: First Half of 1941, January through June.* Washington, DC: USTB.
———. 1941c. *Negro Hotels and Guest Houses, 1941.* Washington, DC: USTB.
———. 1942. *Calendar of Events: Jan.–June 1942.* Washington, DC: USTB.
US Travel Division. 1948. *The Policy and Program of the United States Travel Division*, typescript, mimeograph. Washington, DC: GPO.
US Travel Division. 1949a. *U.S. Travel: A Digest.* Washington, DC: National Park Service, Department of the Interior.
———. 1949b. *National Calendar of Events.* Washington, DC: National Park Service, Department of the Interior.
Velonis, Anthony. 1940. *Technical Problems of the Artist: Technique of the Silk Screen Process.* New York: US Work Projects Administration (WPA), Arts Program.

Secondary Sources

Airey, David. 1984. "Tourism Administration in the USA," *Tourism Management* 5, no. 4 (December): 269–79.
Apostle, Alisa. 2001. "The Display of a Tourist Nation: Canada in Government Film, 1945–1959." *Journal of the Canadian Historical Association* 12, no. 1: 177–97.
Armstead, Myra B. Young. 2005. "Revisiting Hotels and Other Lodgings: American Tourist Spaces through the Lens of Black Pleasure-Travelers, 1880–1950," *Journal of Decorative and Propaganda Arts* 25: 136–59.
Baranowski, Shelley. 2004. *Strength through Joy: Consumerism and Mass Tourism in the Third Reich.* Cambridge, UK: Cambridge University Press.
———, and Ellen Furlough, eds. 2001. *Being Elsewhere: Tourism, Consumer Culture, and Identity in Modern Europe and North America.* Ann Arbor: University of Michigan Press.
Barnes, Harry Elmer, and Oreen M. Ruedi. 1942. *The American Way of Life: Our Institutional Patterns and Social Problems.* New York: Prentice-Hall.
Battaglio, R. Paul, and Jeremy L. Hall. 2018. "Trinity Is Still My Name: Renewed Appreciation for Triangulation and Methodological Diversity in Public Administration" (editorial). *Public Administration Review* 78, no. 6 (November–December): 825–27.

Berger, Dina. 2006. *The Development of Mexico's Tourism Industry: Pyramids by Day, Martinis by Night.* New York: Palgrave Macmillan.
Berkowitz, Michael. 2001. "A 'New Deal' for Leisure: Making Mass Tourism during the Great Depression." In *Being Elsewhere: Tourism, Consumer Culture, and Identity in Modern Europe and North America,* edited by Shelley Baranowski and Ellen Furlough, chap. 8. Ann Arbor: University of Michigan Press.
Bertelli, Anthony M., and Laurence E. Lynn Jr. 2006. *Madison's Managers: Public Administration and the Constitution.* Baltimore: Johns Hopkins University Press.
Boyd, Anne Morris. 1941. *United States Government Publications: Sources of Information for Libraries.* New York: H. W. Wilson.
Brands, H. W. 2008. *Traitor to His Class: The Privileged Life and Radical Presidency of Franklin Delano Roosevelt.* New York: Doubleday.
Brewton, Charles, and Glenn Withiam. 1998. "United States Tourism Policy: Alive, but Not Well." *Cornell Hotel and Restaurant Administration Quarterly* 39, no. 1 (February): 50–59.
Brill, Steven. 2018. *Tailspin: The People and Forces behind America's Fifty-Year Fall— And Those Fighting to Reverse It.* New York: Alfred A. Knopf.
Brinkley, Douglas. 2016. *Rightful Heritage: Franklin D. Roosevelt and the Land of America.* New York: HarperCollins.
Butterfield, Ben. 1970a. "Today's Environmental Challenge." *Environmental Education* 1, no. 4 (Summer): 113–15.
———. 1970b. "The Emerging Leadership of Parks in Today's Travel Environment." *Proceedings of the First Annual Conference, The Travel Research Association, Monterey, Calif., August 16–19, 1970,* 13–17.
Buzard, James. 2001. "Culture for Export: Tourism and Autoethnography in Postwar Britain." In *Being Elsewhere: Tourism, Consumer Culture, and Identity in Modern Europe and North America,* edited by Shelley Baranowski and Ellen Furlough, chap. 12. Ann Arbor: University of Michigan Press.
Carpenter, Daniel P. 2001. *The Forging of Bureaucratic Autonomy: Reputations, Networks, and Policy Innovation in Executive Agencies, 1862–1928.* Princeton, NJ: Princeton University Press.
———. 2010. *Reputation and Power: Organizational Image and Pharmaceutical Regulation at the FDA.* Princeton, NJ: Princeton University Press.
Clark, Christopher. 2013. *The Sleepwalkers: How Europe Went to War in 1914.* New York: HarperCollins.
Cocks, Catherine. 2001. "The Chamber of Commerce's Carnival: City Festivals and Urban Tourism in the United States, 1890–1915." In *Being Elsewhere: Tourism, Consumer Culture, and Identity in Modern Europe and North America,* edited by Shelley Baranowski and Ellen Furlough, chap. 4. Ann Arbor: University of Michigan Press.

Cook, Brian J. 2014. *Bureaucracy and Self-Government: Reconsidering the Role of Public Administration in American Politics*, 2nd ed. Baltimore: Johns Hopkins University Press.

Dawson, Michael. 2004. *Selling British Columbia: Tourism and Consumer Culture, 1890–1970*. Vancouver, BC: UBC [University of British Columbia] Press.

———. 2011. "'Travel Strengthens America'? Tourism Promotion in the United States during the Second World War." *Journal of Tourism History* 3, no. 3 (November): 217–36.

Donovan, Frank, Jr. 1940. "The Railroad in Literature as a Public Relations Medium." *Railway Age* 109, no. 9 (August 31): 307, 312.

Dubinsky, Karen. 1999. *The Second Greatest Disappointment: Honeymooning and Tourism at Niagara Falls*. New Brunswick, NJ: Rutgers University Press.

Duchemin, Michael. 2009. "Water, Power, and Tourism: Hoover Dam and the Making of the New West." *California History* 86, no. 4: 60–78, 87–89.

———. 2016. *New Deal Cowboy: Gene Autry and Public Diplomacy*. Norman: University of Oklahoma Press.

Durant, Robert F. 2014. "Taking Time Seriously: Progressivism, the Business–Social Science Nexus, and the Paradox of American Administrative Reform." *PS: Political Science & Politics* 47, no. 1 (January): 8–18.

Edgell, David L. 1983. "United States International Tourism Policy." *Annals of Tourism Research* 10, no. 3: 427–33.

———. 1984. "US Government Policy on International Tourism." *Tourism Management* 5, no. 1 (March): 67–70.

———. 1992. "United States Travel and Tourism Administration." *Annals of Tourism Research* 19, no. 3: 595–97.

Egdall, David L., Sr., and Jason R. Swanson. 2013. *Tourism Policy and Planning: Yesterday, Today, and Tomorrow*, 2nd ed. New York: Routledge.

———. 2019. *Tourism Policy and Planning: Yesterday, Today, and Tomorrow*, 3rd ed. New York: Routledge.

Fehrenbacher, Don E. (1978) 2001. *The Dred Scott Case: Its Significance in American Law and Politics*. Oxford, UK: Oxford University Press.

Fox, Daniel M. 1961. "The Achievement of the Federal Writers' Project." *American Quarterly* 13, no. 1 (Spring): 3–19.

Furlough, Ellen. 1998. "Making Mass Vacations: Tourism and Consumer Culture in France, 1930s to 1970s." *Comparative Studies in Society and History* 40, no. 2 (April): 247–86.

Gabriele, Kathryn R. 2015. "Lessons from a Buried Past: Settlement Women and Democratically Anchored Governance Networks." *Administration & Society* 47, no. 4: 393–415.

Green, William H., ed. 1940. *The Negro Motorist Green Book, 1940 Edition*. New York: Victor H. Green.

———, ed. 1941. *The Negro Motorist Green Book, 1941 Edition*. New York: Victor H. Green.
Griswold, Wendy. 2016. *American Guides: The Federal Writers' Project and the Casting of American Culture*. Chicago: University of Chicago Press.
Gunn, Clare A. 1983. "U.S. Tourism Policy Development." *Journal of Physical Education, Recreation and Dance* 54, no. 4 (April): 32–35.
Henry, Thomas P. 1941. "The Travel Industry." In *The Development of American Industries: Their Economic Significance*, rev. ed., edited by John George Glover and William Bouck Cornell, chap. 39. New York: Prentice-Hall.
Hoffer, Williamjames Hull. 2007. *To Enlarge the Machinery of Government: Congressional Debates and the Growth of the American State, 1858–1891*. Baltimore: Johns Hopkins University Press.
Jick, Todd D. 1979. "Mixing Qualitative and Quantitative Methods: Triangulation in Action." *Administrative Science Quarterly* 24, no. 4 (December): 602–11.
King, David C. 1994. "The Nature of Congressional Committee Jurisdictions." *American Political Science Review* 88, no. 1 (March): 48–62.
Lambright, W. Henry. 2007. "Leading Change at NASA: The Case of Dan Goldin." *Space Policy* 23, no. 1 (February): 33–43.
———. 2017. "Reflections on Leadership and Its Politics: Charles Bolden, NASA Administrator, 2009–17." *Public Administration Review* 77, no. 4 (July–August): 616–20.
Lee, Mordecai, ed. 2005a. "Symposium on Rewriting the History of Public Administration: What If . . . ," *Public Voices* 8, no. 1: 3–60.
———. 2005b. *The First Presidential Communications Agency: FDR's Office of Government Reports*. Albany: State University of New York Press.
———. 2010. *Nixon's Super-Secretaries: The Last Grand Presidential Reorganization Effort*. College Station: Texas A&M University Press.
———. 2011. *Congress vs. the Bureaucracy: Muzzling Agency Public Relations*. Norman: University of Oklahoma Press.
———. 2012. *Promoting the War Effort: Robert Horton and Federal Propaganda, 1938–1946*. Baton Rouge: Louisiana State University Press.
———. 2014. "Government Is Different: A History of Public Relations in American Public Administration." In *Pathways to Public Relations: Histories of Practice and Profession*, edited by Burton St. John III, Margot Opdycke Lamme, and Jacquie L'Etang, chap. 7. London: Routledge.
———. 2016. *A Presidential Civil Service: FDR's Liaison Office for Personnel Management*. Tuscaloosa: University of Alabama Press.
———. 2017a. "The Practice of Public Affairs in Public Administration." In *SAGE Handbook of International Corporate and Public Affairs*, edited by Phil Harris and Craig S. Fleisher, chap. 12. Thousand Oaks, CA: Sage.

———. 2017b. "Trying to Professionalize Expert Knowledge, Part II: A Short History of Public Administration Service, 1933–2003." *Public Voices* 15, no. 1: 28–45.
———. 2018. *Get Things Moving! FDR, Wayne Coy, and the Office for Emergency Management, 1941–1943*. Albany: State University of New York Press.
———. 2019. "Revitalizing historiography in public administration." *Public Performance and Management Review*, DOI: 10.1080/15309576.2019.1677256.
———, Fraser Likely, and Jean Valin. 2017. "Government Public Relations in Canada and the United States." In *North American Perspectives on the Development of Public Relations*, edited by Tom Watson, chap. 6. London: Palgrave Macmillan/Springer.
Lockwood, John E., and Luther Ely Smith, Jr. 1941. "Intra-Governmental Activities of the United States to Foster Hemispheric Trade." *Law and Contemporary Problems* 8, no. 4 (Autumn): 669–83. (Note: Notwithstanding the listed season of publication, this issue came out after Pearl Harbor.)
Louter, David. 2001. "Glaciers and Gasoline: The Making of a Windshield Wilderness, 1900–1915." In *Seeing and Being Seen: Tourism in the American West*, edited by David M. Wrobel and Patrick T. Long, chap. 11. Lawrence: University Press of Kansas.
Lucander, David. 2014. *Winning the War for Democracy: The March on Washington Movement, 1941–1945*. Urbana: University of Illinois Press.
Mak, James. 2015. "Creating 'Paradise of the Pacific': How Tourism Began in Hawaii," monograph. Working Paper No. 2015-1. Economic Research Organization, University of Hawaii, Honolulu.
Mashaw, Jerry L. 2012. *Creating the Administrative Constitution: The Lost One Hundred Years of American Administrative Law*. New Haven, CT: Yale University Press.
McDonald, Bruce D., III. 2010. "The Bureau of Municipal Research and the Development of a Professional Public Service." *Administration & Society* 42, no. 7: 815–35.
McDowell, Chas. A. R. 1940. "Southward." In *The Negro Motorist Green Book, 1940 Edition*, edited by William H. Green, 26–27, 31. New York: Victor H. Green.
McGarry, William A. 1936. "Going Places." *Rockefeller Center Weekly* 5, no. 10 (September 4): 4, 15.
McLennan, Sarah Elizabeth. 2015. "Promoting Tourism, Selling a Nation: The Politics of Representing National Identity in the United States, 1930–1960." Unpublished dissertation for Ph.D. in history, College of William and Mary, Williamsburg, VA.
McNabb, David E. 2018. *Research Methods for Public Administration and Nonprofit Management: Quantitative and Qualitative Approaches*, 4th ed. New York: Routledge.
Meyers, Roy T., and Irene S. Rubin. 2011. "The Executive Budget in the Federal Government: The First Century and Beyond." *Public Administration Review* 71, no. 3 (May–June): 334–44.

Nathan, Richard P. 1975. *The Plot that Failed: Nixon and the Administrative Presidency*. New York: John Wiley & Sons.
——. (1983) 1986. *The Administrative Presidency*. New York: Macmillan.
Newbold, Stephanie P. 2010. *All but Forgotten: Thomas Jefferson and the Development of Public Administration*. Albany: State University of New York Press.
——, and Larry D. Terry. 2006. "The President's Committee on Administrative Management: The Untold Story and the Federalist Connection." *Administration & Society* 38, no. 5 (November): 522–55.
——, and David H. Rosenbloom, eds. 2007. Symposium on Brownlow Report Retrospective. *Public Administration Review* 67, no. 6 (November–December): 1006–58.
Ogg, Frederick A. 1935. "Thirtieth Annual Meeting of the American Political Science Association." *American Political Science Review* 29, no. 1 (February): 107–13.
Pautler, Paul A. 2015. "A Brief History of the FTC's Bureau of Economics: Reports, Mergers, and Information Regulation." *Review of Industrial Organization* 46, no. 1 (February): 59–94.
Pfiffner, James P., and Douglas A. Brook, eds. 2000. *The Future of Merit: Twenty Years after the Civil Service Reform Act*. Washington, DC and Baltimore: Woodrow Wilson Center Press and Johns Hopkins University Press.
Pillen, Cory. 2008. "See America: WPA Posters and the Mapping of a New Deal Democracy." *Journal of American Culture* 31, no. 1 (March): 49–65.
——. 2020. *WPA Posters in an Aesthetic, Social, and Political Context: A New Deal for Design*. New York: Routledge.
Popp, Richard K. 2012. *The Holiday Makers: Magazines, Advertising, and Mass Tourism in Postwar America*. Baton Rouge: Louisiana State University Press.
Raadschelders, Jos C. N. 2017. "Administrative History as a Core Element in the Study of Public Administration." In *Foundations of Public Administration*, edited by Jos C. N. Raadschelders and Richard J. Stillman II, chap. 3. Irvine, CA: Melvin & Leigh.
Reed, Nathaniel Pryor. 2016. *Travels on the Green Highway: An Environmentalist's Journey*. Hobe Sound, FL: Reed Publishing.
Rockwell, Stephen J. 2010. *Indian Affairs and the Administrative State in the Nineteenth Century*. New York: Cambridge University Press.
Rosenbloom, David H. 2000. *Building a Legislative-centered Public Administration: Congress and the Administrative State, 1946–1999*. Tuscaloosa: University of Alabama Press.
Santoro, Vincenzina. 1967. "Tourism and the United States Balance of Payments." *Tourist Review* 22, no. 1: 28–32.
Schachter, Hindy Lauer. 2017. "Organization Development and Management History: A Tale of Changing Seasons." *Public Administration Quarterly* 41, no. 2 (Summer): 233–53.

Sellars, Richard West. 1993. "The Rise and Decline of Ecological Attitudes in National Park Management, 1929–1940; Part III (Conclusion): Growth and Diversification of the National Park Service." *George Wright Forum* 10, no. 3: 38–54.

Selznick, Philip. (1949) 1984. *TVA and the Grass Roots*. Berkeley: University of California Press.

Semmens, Kristin. 2005. *Seeing Hitler's Germany: Tourism in the Third Reich*. New York: Palgrave Macmillan.

Shaffer, Marguerite S. 2001. *See America First: Tourism and National Identity, 1880–1940*. Washington, DC: Smithsonian Institution Press.

Sklaroff, Lauren Rebecca. 2009. *Black Culture and the New Deal: The Quest for Civil Rights in the Roosevelt Era*. Chapel Hill: University of North Carolina Press.

Sorin, Gretchen Sullivan. 2009. "'Keep Going': African Americans on the Road in the Era of Jim Crow." Unpublished dissertation for Ph.D. in history, State University of New York-Albany.

Steil, Benn. 2018. *The Marshall Plan: Dawn of the Cold War*. New York: Simon & Schuster.

Stivers, Camilla. 2000. *Bureau Men, Settlement Women: Constructing Public Administration in the Progressive Era*. Lawrence: University Press of Kansas.

———. 2011. "Administration and the Limits of Democracy: The Space of 19th-Century American Governance." *Administration & Society* 43, no. 6: 623–42.

Swain, Donald C. 1972. "The National Park Service and the New Deal, 1933–1940." *Pacific Historical Review* 41, no. 3 (August): 312–32.

Thornton, Robert L. 1969. "Tourism—West from Europe: The Enticement of the Foreign Visitor." *Business Horizons* 12, no. 5 (October): 27–34.

U.S. Geological Survey: Its History, Activities and Organization. 1918. New York: D. Appleton.

van Thiel, Sandra. 2014. *Research Methods in Public Administration and Public Management: An Introduction*. London: Routledge.

Weber, Gustavus A., and Laurence F. Schmeckebier. 1934. *The Veterans Administration: Its History, Activities and Organization*. Washington, DC: Brookings Institution.

White, Richard D. 2003. *Roosevelt the Reformer: Theodore Roosevelt as Civil Service Commissioner, 1889–1895*. Tuscaloosa: University of Alabama Press.

Williams, Daniel W. 2002. "Before Performance Measurement." *Administrative Theory & Praxis* 24, no. 3: 457–86.

Williams, Wythe. 1938. "Uncle Sam Invites Europe to Call." *Commentator* 2, no. 6 (January): 82–86.

Wilson, James Q. 2000. *Bureaucracy: What Government Agencies Do and Why They Do It*, new ed. New York: Basic Books/Perseus.

Wrobel, David M., and Patrick T. Long. 2001. *Seeing and Being Seen: Tourism in the American West*. Lawrence: University Press of Kansas.

Zavattaro, Staci M. 2015. "Place Brand Identity: An Exploratory Analysis of Three Deep South States." *International Journal of Organization Theory and Behavior* 18, no. 4 (Winter): 405–432.

———. 2018. "What's in a Symbol? Big Questions for Place Branding in Public Administration." *Public Administration Quarterly* 42, no. 1 (Spring) 90–119.

———, and Daniel L. Fay. 2019. "Brand USA: A Natural Quasi-Experiment Evaluating the Success of a National Marketing Campaign." *Tourism Management* 70 (February): 42–48.

Zelizer, Julian E. 2012. *Governing America: The Revival of Political History*. Princeton, NJ: Princeton University Press.

Index

Action, 130, 195n66
advisory boards
 business interests and, 104, 121–122
 Office of Travel, 121–123
 private sector and, 133
 public interest and, 151
 USTD, 104–105, 106, 108
African Americans. *See also* racism; segregation
 civil rights of, x, 32, 144–145
 discriminatory hiring practices, x, 145
 Esso gas station promotion, 73
 Jackson State University shooting, 193n53
 outreach to, 32–33
 racism and, 11
 recreational needs of, x, 73
 Republican party and, 32
 Roosevelt's attitudes toward, x, 32, 144–145
 segregation of, 42, 43, 72, 146–147, 176n48
 sensitivity toward needs of, 42, 146–147
 "separate but equal" treatment of, 146–147
 southern places of interest to, 56
 travel agencies catering to, 43
 travel brochures, 168n59
 travel clubs for, 41, 43
 travel promotion for, x–xi, 11, 32–33, 41–43, 56–57, 72–74, 144–147
 USTB staff, xi, 32–33, 41, 146
Afro-American, 73
Afro-American Life Insurance Company, 43
Airey, David, 6, 115, 133
airlines
 promotional information from, 5
 promotion of, 55
 reservations, 76, 177n64
 support of USTB by, xi, 35
 wartime travel, 88
Air Transport Association of America, 44, 109
Alaska, 122–123, 125
Allen, Horton S., 69, 160n8
American Automobile Association (AAA), 36, 66, 72, 77, 79, 86, 87
American Express Company, 25, 159n26
American Guide Series, 54
American Hotel Association, 17, 29, 44, 68, 143–144, 164n9
American Petroleum Institute, 121
American Political Science Association, 165n20

215

Index

American Recreation Association, 78–79
American Society for Public Administration (ASPA), 156n7
American Steamship Owners' Association, 18
American territories, 40
American Travel Development Association, 16
American West, 28, 71
Anderson, J. R., 30
Apostle, Alisa, 3
appropriation legislation, 9, 155n3
Arches National Park poster, x, following page 62
Armstead, Myra B. Young, 3, 42, 74
Army camp V-days, 79
artists, ix, 22
Associated Negro Press (ANP), 73, 74
Associated Press (AP), 96, 188–189n39
Atlantic City (NJ) Chamber of Commerce, 90
austerity effort, 18
authorization legislation, 8–9, 155n3
auto industry, 29
Autry, Gene, 3

Bailey, E. J., 172n48
Bailey, Josiah, 43, 44, 58–60, 168n70
Baltimore American, 176n48
Baltimore & Ohio (B&O) Railroad, 95, 183n70
Baranowski, Shelley, 3, 10
Barkley, Alben, 60
Barnes, Harry Elmer, 70
Barron's, 50
Battaglio, R. Paul, 12
Battle of Britain, 49, 61
beautification, 106
Bennett, Philip, 92
Berger, Dina, 3, 7
Berkowitz, Michael, 3, 7
Bernays, Edward, 165n20

Bible, Alan, 125
Bilbo, Theodore, 59–61
Blackwell, Jean, *following p. 62*
BLM (Bureau of Land Management), 119
Board of Economic Warfare, 83–84, 179n7
Bossemeyer, J. L., 188–189n47, 188n26
  appointed USTB director, 105
  full name, 160n12
  promotion of tourism by, 28, 71
  travel statistics and, 106
  USTB closing announced by, 111
  USTB promoted by, 30, 107
  USTB reopening and, 105
  USTB reorganization and, 69
  USTB wartime restructuring and, 88, 89
  on wartime tourism, 84
Boston, Rosalind, 41
Boston Globe, 27–28, 102
Boulder (Hoover) Dam, 36, 138
Boyde, Anne Morris, 37
Brands, H. W., xi
Brand USA, 134
Bratter, Herbert M., 16
Brewton, Charles, 7
Brill, Steven, 150
Brinkley, Douglas, 11, 20, 141
brochures, 40, 41, 51
Brook, Douglas A., 2
Brookings Institution, 3
Brownlow, Louis, 165n20
Brownlow Committee report, 2–3
Bryan, William Jennings, 37
Bureau of Colored Work of the National Recreation Association, 73
Bureau of Foreign and Domestic Commerce, Commerce Department, 15, 16, 44–45
  Tourist Travel Division proposed in, 18–19
Bureau of Indian Affairs, 3, 119

Bureau of Land Management (BLM), 119, 123, 138
Bureau of Mines and Solid Fuel Administration, 95, 105
Bureau of Outdoor Recreation, 119
Bureau of Reclamation, 105, 119, 136, 138
Bureau of the Budget (BOB), 136, 169n73, 172n47, 172n48, 184n80
  approval of open-ended cap by, 125
  NPS Travel Division proposal and, 17–18
  opposition to USTB by, 34
  reprogramming and, 118
  revisions requested by, 59
  USTB closing and, 96–97
  USTB funding and, 45
  USTB refunding recommended by, 102–103
Burlew, Ebert K., 172n48
Burtness, Olger, 148
business organizations. *See also* private-sector interests; travel industry
  advisory committees, 104–105, 106, 108, 121–122, 151
  calls for government travel office by, 113
  cooptation of federal government with, 57–58, 143–144, 151
  lobbying by, 38
  Office of Travel advisory board, 121–123
  Republican party and, 103–104
  support of USTB by, xi–xii, 4, 35–37, 43–44, 50, 58, 104
  tourism and, 15–17
  USTD and, 104–105, 107–108
Business Week, 18–19
Butterfield, Alexander, 196n72
Butterfield, Ben
  Alaska publications, 123
  appointed USTB director, 120–121
  national park promotion and, 131
  political power struggles and, 124, 126
  travel themes, 124
Butz, Earl, 195n70
Buzard, James, 10
Byrnes, James, 162n40

Calendar of Events, 40, 70, 88
Cambodian invasion, 127
Cammerer, Arno, 27, 32, 68, 159n26
Canada
  coordination with, 50, 52
  government travel promotion in, 10, 100, 189n45
  youth travel program, 196n79
Caribbean, 29, 35, 74
Carlsbad Caverns National Park poster, x, *following p. 62*
Carpenter, Daniel P., 3, 6
Carter, Albert, 38, 67, 99
Carter, Jimmy, 133, 149, 151
CCC (Civilian Conservation Corps), 22, 23, 51
Census Bureau, 106
Central America, 52, 141. *See also* Latin America
chambers of commerce
  Atlantic City (NJ), 90
  New York State, 157n3
  support of USTB by, xi
  tourism promotion by, 1
  United States, 15, 44, 103, 143
  Virginia, 70
Chapman, Oscar, 29
Charlestown [WV] Gazette, 22
Chicago Tribune, 39, 108, 111
Christian Science Monitor, 37, 75, 84, 91, 100
Churchill, Winston, 169n77
Civil Aeronautics Authority (CAA), 44, 55
Civilian Conservation Corps (CCC), 22, 23, 51

civil rights. *See also* African Americans; racism
  opposition to, x, 32
  Roosevelt, Eleanor and, 144–145
  Roosevelt's reluctance to support, x, 32, 144–145
Clark, Bennett "Champ," 60–61
Clark, Christopher, 13–14, 85
Clawson, Ken W., 194*n*59
Cleveland Plain Dealer, 86
Clinton, Bill, 133
Cocks, Catherine, 1
collaborators
  Gerard, 23, 24, 28
  Krug's appointment of, 105
  McDowell, 32, 56, 72
  Rohde, 37, 166*n*31
  use of term, 158*n*12
Collins, James, 149
Colored Professional and Business Men's Club, 42
color slides, 69
Commentator, 31
Commerce Department
  Bureau of Foreign and Domestic Commerce, 15, 16, 18–19, 44–45
  domestic travel promotion under, 7, 14, 123, 129–132
  National Travel and Tourism Office, 134
  Office of International Travel, 114
  Office of Travel Promotion, 133–134
  opposition to USTD/USTB by, 109, 169*n*73
  public relations use by, 4
  support of USTB by, 47
  Tourism Policy Council, 133
  tourism promotion by, 1, 20, 124–127
  turf battles over travel bureau, 7, 33–34, 102, 106, 113, 137–138, 169*n*76, 187*n*13, 194*n*63
  United States Travel Service (USTS), 115, 120, 123, 124–126, 133
  USTB funding process and, 44–47
  USTB transferred to, 7, 138
  USTD and, 111
  US Travel and Tourism Administration and, 133
Community Development Department, 128, 194*n*61
Congress, US. *See also* House Appropriations Committee; House Commerce Committee; House Interstate and Foreign Commerce Committee; Senate Appropriations Committee; Senate Commerce Committee; Senate Interior Committee
  appropriations bills, 9, 155*n*3
  appropriations process, 64–68, 102–104, 107–111
  authorization bills, 8–9
  authorization process, 57–61
  budgeting process, 65–66
  Eightieth Congress, 103–104, 107–110
  Eighty-First Congress, 110–111
  opposition to public relations by, xii, 5–6, 58
  prohibition of lobbying, 165*n*22
  Seventy-Fifth Congress, 21–22, 33–34, 58
  Seventy-First Congress, 15–16
  Seventy-Fourth Congress, 17–20
  Seventy-Ninth Congress, 102–103
  Seventy-Second Congress, 16–17
  Seventy-Seventh Congress, 64–68

Seventy-Sixth Congress, 1–2, 4, 43–48, 57–61
Seventy-Third Congress, 17
Congressional Directory, 88, 89
Congressional Record, 57
Congressional Travel & Tourism Caucus, 133
congressional veto, 130, 194–195*n*64
conservation. *See also* environmental protection
  tourism and, 141
  Travel America and, 50
  USTB mission and, xii, 11–12
  USTD mission and, 12, 106
conservatives
  Garner and, 186*n*4
  opposition to New Deal programs by, 57
  opposition to public relations by, xii, 58
  on proper role of government, 148
  support of USTB by, 30–31, 44
  USTB public relations role and, 137
  wartime USTB budget discussions and, 91–92
Constitution, US, 64, 65, 94, 147
consumer spending, 83
Cook, Brian, 144
cooptation, 57–58, 143–144, 151
Copeland, Royal, 17, 18, 19, 34, 163*n*62
Coy, Wayne, 147, 156*n*7, 162*n*40, 184*n*80
Cuba, 101

dams, 4, 36, 92, 119, 138
Dannemeyer, William, 149
Davis, Jefferson, 168*n*59
Dawson, Michael, 3, 10
defense industry categorization, 177*n*56
defense mobilization, 177*n*56

defense workers, 82
Demaray, Arthur E., 38
  promotion of USTB by, 31–32
  USTB budget hearings and, 66
  USTB creation and, 21–22
  USTB funding and, 102–103
  wartime USTB funding and, 92
Democratic party
  opposition to USTD by, 110–111
  USTB funding and, 91–92
  USTB public relations and, 137
Dent, Frederick, 195*n*70
Dewey, Thomas E., 166*n*23
Dimock, Marshall, 165*n*20
Directory of Negro Hotels and Guest Houses in the United States (USTB), xi, 42, 56. See also *Green Book, The*
Discover America campaign, 117, 118, 120
Division of Information, Interior Department, 137
Division of Negro Affairs, ix, 42, 72, 146
Division of Tourism, National Park Service (NPS), 120–121
Dixie to the New York World's Fair, 42
Dolan, D. Leo, 170*n*4
domestic tourism. *See also* foreign tourism; tourism
  by African Americans, x–xi, 11, 32–33, 41–43, 56–57, 72–74, 144–147
  benefits of, 20
  Interior Department and, 8
  patriotism and, xiii, 15, 30, 77, 79, 91, 92, 142, 5075
  trade deficit and, 116
  USTB promotion of, x, xii, 1, 6–7, 10–12
  USTD focus on, 105

domestic tourism *(continued)*
  in wartime, 81–85, 141–142
  World War II and, 48, 49–50
Donavan, Frank, Jr., 55
Dorsett, Harold L., 29, 55
Dressler, Marie, 42
Drury, Newton B., 92, 98, 182*n*53
Dubinsky, Karen, 10
DuBois, W. E. B., 57
Duchemin, Michael, 3
dude ranches, 28, 43, 90, 140, 148
Durant, Robert F., 2
Dux, Alexander, *following p.* 62
Dyer, Leonidas, 15–17

Eastern Travel Today, 73
ECA (Economic Cooperation Administration), 113
ecology tourism, 139
Economic Affairs Department, 128
Economic Cooperation Administration (ECA), 113
Economic Defense Board, 179*n*7
economy bloc, 91–92, 94
ECW (Emergency Conservation Work) Act of 1933, 22
Edgell, David L., 6, 133
Edwards, Max N., 122
Ehrlichman, John, 195*n*70
Eightieth Congress, 103–104, 107–110
Eighty-First Congress, 110–111
Eisenhower, Dwight D., 114, 178*n*69
Emergency Conservation Work (ECW) Act of 1933, 22
Emergency Relief Act (ERA), 39–40
employment
  of African Americans, x, 145
  of artists, ix
  federal policies and, 140, 145
  in government public relations, 10
  military services, 95–96

  New Deal and, 17, 18
  nondiscrimination in, x, 145
  private sector and, 145
  statistics, 5, 152
  in travel and tourism, 150
  unemployment, 18, 50
  wartime expenditures and, 50
engineering marvels, 36, 92
environmental movement, 12
environmental protection. *See also* conservation
  ecology tourism and, 139
  speeches about, 124
  USTB mission and, 11–12
  USTD and, 106
Environmental Protection Agency (EPA), 12, 130
Esso gas stations, 73
Europe
  American tourism in, 113–114
  grand tours, 141
  leather goods imported from, 71
European tourists
  attraction of, 30, 31, 40
  German Jews, 156*n*9, 163*n*66, 168*n*70, 173*n*55
  opposition to, 19–20
  postwar attraction of, 114–115
  reduced focus on, 105
  in wartime, 24, 35, 36, 49, 141–142
executive branch reorganization, 128–132
Executive Office of the President, 63, 95, 155*n*3

Fair Employment Practices Committee (FEPC), x
Fay, Daniel L., 134
Federal Art Project, 53, *following p.* 62
Federal Emergency Relief Administration, 47

federal government
  closing federal agencies, 97
  cooptation with travel industry, 143–144, 151
  efforts to reign in, 19–20, 97, 103
  expansion of role of, 4, 19–20, 29, 30, 57
  information disseminated by agencies of, 4–5
  official publications, 13
  proper role in travel and tourism, 147–151
  public relations use by, 1, 4–6, 10–11, 58
  Republican views on, 19–20
  studies of specific agencies, 2–3
  tourism and, 15–26
  travel promotion by, 1–22, 4–6
  wartime efforts to reduce size of, 91–92
federal holidays, 71
Federal Security Agency (FSA), 46, 76, 178n69
Federal Trade Commission (FTC), 3
Federal Writers' Project (FWP), 22, 40, 54
Fehrenbacher, Don E., 150
FEPC. *See* Fair Employment Practices Committee (FEPC)
films
  distribution of, 41, 54, 69, 98
  military services and, 95
  Spanish and Portuguese translations, 52, 68, 89, 141
  travelogues, 68, 69
  USTB budget for, 60
Fish and Wildlife Service, 106, 119, 138
Floethe, Richard, 53
Florida, 30, 36, 43
Food and Drug Administration, 3
Ford, Gerald, 195n70

foreign tourism. *See also* domestic tourism; tourism
  benefits of, 15, 24, 116–118
  Brand USA and, 134
  development of, 22–25, 29–30, 67, 113–116
  encouraging, 5, 6, 19, 20, 29, 30, 40, 105, 113–116, 118, 138, 156n9
  to national parks, 115–116, 118
  promotional materials for, 5, 25, 29, 40, 105
  State department concerns about Jewish tourists, 156n9, 163n66, 168n70, 173n55
  taxing, 151
Forest Service, USDA, 106
Fox, Daniel M., 54
Fresno [CA] Bee, 44
Furlough, Ellen, 3, 10

Gabriele, Kathryn R., 2
Garner, John Nance, 186n4
gasoline supply, 77–78, 85, 86, 178n69, 181n23
Gaus, John, 165n20
General Land Office, 3
Gerard, James, 157n19, 161n33
  as collaborator to USTB, 23–24, 32
  honorary president of National Resorts and Parks Associations, 20
  Ickes and, 160n9
  industry support for USTB and, 33–34
  promotion of USTB by, 30
  resignation of, 28, 37
  role of, 24
German, Anna, 41
Germany
  aggression by, 49, 63, 173n55
  Axis powers and, 170n3

Germany *(continued)*
  Latin America and, 141
  submarine warfare, 64, 170n5
  tourism promoted in, 10, 82
  US concerns about immigration of Jews from, 156n9, 163n66, 168n70, 173n55
  US relations with, 64
Gettysburg National Park, 116
Glasgow, Leslie, 127, 194n58
Good Neighbor Policy, 24, 29, 35, 49
Graham, Katherine, 156n7
Graham, Philip L., 155–156n7
Grand Canyon, 116
graphic artists, ix–x
Grazing Service, Interior Department, 55
Great Britain, 50
Great Depression
  New Deal and, 17
  tourism promotion and, 140
  travel spending and, xii
  WPA support of artists during, ix
Great Recession, 179n93
Greeley, Horace, 26, 159n8, 159n28
Green, Victor, 146
Green, William H., xi, 56, 72
Green Book, The, 11, 56, 72, 176n48. See also *Directory of Negro Hotels and Guest Houses in the United States* (USTB); *Negro Motorist Green Book, The* (Green)
Griswold, Wendy, 3, 54
Gunn, Clare A., 7, 133

Hall, Jeremy L., 12
Halls, Richard, following p. 62
Hartzog, George B., Jr., 13, 195–196n71
  free media publicity and, 123–124
  Johnson's decision not to run and, 119–122

  Morton's firing of, 131
  national parks tourism and, 118
  political power struggles and, 124–127
  "See America" campaign and, 116
  USTB funding and, 119–120, 191n20, 191n21
  USTB legislation and, 193n46
Hatch Act, 61
Hawaii, 116, 125, 132
Hayden, Carl, 67–68, 94
Henni, Clara, 95, 184n88
Henry, Thomas, 72
Herald Tribune, 61, 75, 91, 173n61, 175n30
Herring, Pendleton, 165n20
Herzog, Harry, *following p.* 62
Heselton, John, 149
Heuser, Mathilda C., 95, 184n88
Hickel, Walter, 193n41, 194n55, 196n74
  Nixon's firing of, 127
  political power struggles and, 126, 127
  USTB under, 122–123
historical evaluation standard, 13–14
historical sites, 54
Hitler, Adolf, 35, 48, 63, 141, 156n9, 168n70, 169n77
Holland, Spessard, 181n23
Hoover, Herbert, 16
Hoover Commission, 110
Hoover Dam, 36, 119, 138
Hopkins, Harry, 169n76, 169n77
  appointed commerce secretary, 46
  on artists, ix
  illness of, 46
  USTB creation and, 1
  USTB jurisdictional dispute and, 46–47
  USTB structure and, 138
Horton, Robert, 36, 170n5

hospitality industry, 83
hotels
  fixed costs of, 83
  racial segregation in, 176n48
  segregation in, 176n48
  wartime availability of, 36, 83, 86, 91
  welcoming to African Americans, xi, 11, 41–43, 72, 74
House Appropriations Committee, 9, 64–68, 67, 68, 92, 102, 103, 110, 125, 149
  Interior Subcommittee, 98
House Commerce Committee, 9, 15, 17, 20, 34, 43, 44, 47, 58, 125, 126, 192n32
House Interstate and Foreign Commerce Committee, 15, 17, 20, 109, 117
Hull, Cordell, 18, 168n70
human resources, 194n60
Human Resources Department, 128
Humphrey, Hubert, 116–117, 119
Hundred Days, New Deal, 17

Ickes, Harold, 169n76, 170n5, 172n47, 173n61
  civil rights and, 32, 146
  criticism of, 137
  diary, 37, 38, 46, 87, 181n35
  Gerard and, 160n9
  goals for USTB, 26
  industry support and, 35–37, 151
  legislative lobbying by, 33–34
  Loomis fired by, 38, 39
  Macnamee and, 165n20
  media and, 70
  national parks visits and, 20
  as petroleum coordinator, 77–78, 86–88
  political constraints on, 51, 52, 57
  responsibilities of, 162n40, 178n74
  Rohde and, 37, 38–39, 165n14
  speech to AAA, 36–37
  tourism promoted by, 29–30
  Truman and, 102
  USTB budget and, 25, 66–67
  USTB created by, 1–2, 21–22, 132, 135
  USTB legislation and, 43–47, 58–61, 157n1, 164n9
  USTB promoted by, 31
  USTB structure and, 7, 57, 105, 113, 130, 138
  USTB wartime reductions and, 91, 93
  US Travel Commission and, 17
  wartime criticism of, 86–88
  wartime public relations and, 90
  wartime tourism and, 85–88, 181n23
Ickes Papers, 13
Industry-Government Special Task Force on Travel, 117
industry lobbying, 38, 143–144
inflation, 83
information dissemination, 4–5. *See also* promotional materials; public relations (PR)
Inouye, Daniel, 126–127, 132, 196n78
Institute for Government Research, 3
inter-American tourism, 35–36, 52, 57
Interior Department. *See also* National Park Service (NPS); United States Travel Bureau (USTB)
  African American needs examined by, x
  appropriation of funds for, 9
  authorization of funds to, 9
  Division of Information, 137
  Johnson's promotion of domestic travel by, 115–122
  Library, 13
  Nixon's reorganization proposals and, 195n70

Interior Department *(continued)*
  Office of Education, 41
  Office of Travel and Information Services, 14, 119–122
  postwar tourism promotion and, 115–122
  public relations use in, xii
  purpose of, 21
  role in tourism promotion, 124–127
  secretarial orders, 1, 12–13
  Secretary's Order 2912, 120–121
  transfer of USTB to Office of the Secretary, 37–38, 51
  turf battles over USTB, 137–138, 187n13, 194n63
  US Geological Survey, 3
  USTB in, 1, 7, 8, 33–34, 45, 61
International Trade Federation, 16
International Travel Act of 1961, 115
International Travel Federation, 15, 156n3
Interstate and Foreign Commerce Committee, 109
Irene Mound, 56
isolationism, 142
It's Easier than You Think to Vacation in Alaska This Spring, 123

Jackson, Henry "Scoop," 120
Jackson, James A., 73
Jackson State University, 193n53
Jensen, Ben, 108
Jewish German tourists, 156n9, 163n66, 168n70, 173n55
Jick, Todd D., 12
Johnson, Eugene I., 55
Johnson, Jed, 93, 98–99
Johnson, Lyndon B., 191n18, 191n19
  announces he will not run for reelection, 118
  federal budget, 118
  reelection campaign, 191n18, 191n24
  travel task force, 116, 117
  USTB and, 10, 12, 14, 115–120, 122, 125
Jones, Robert, 66, 99, 149
journalism, 13
Journal of Adult Education, 31

Kaltenborn, H. V., 163n50
Kennedy, John F., 6, 115
Kennedy, Robert F., 119
Kent State University, 127
King, David C., 3, 9
Korean War, 193n42
Krug, Julius A.
  reestablishment of USTB by, 104–105
  USTB funding and, 102–103, 107–108, 110–111
Kucera, Beth, xiii

Lambright, W. Henry, 3
Las Vegas, 179n93
Latin America, 30, 35, 40, 49–50, 52, 67, 69, 74, 79, 89, 95, 105
  German activity in, 141
  promoting closer relations with, 170n3
  Roosevelt's pursuit of closer ties with, 141
  USTB promotion of travel in, 6, 7
Lattimore, Ralston B., 188n29
  USTB promotional activities, 107
Lea, Clarence, 17–20, 19, 43, 93, 109
Leavy, Charles, 66
Lee, Mordecai, x, xii, 1, 2, 4, 6, 10, 12, 30, 36, 47, 58, 63, 100, 104, 128, 130, 131, 137, 140, 145, 147, 155n3
legislative veto, 130, 194–195n64
Lend-Lease, 50, 63, 169n77
Likely, Fraser, 10

limited national emergency declaration, 50, 64, 142, 173n2
lobbying, 33–34, 38, 143–144, 165n22
Lockwood, John E., 89
Long, Breckenridge, 173n55
Long, Patrick T., 3
Long, Russell B., 197n5
Loomis, Nelson W., 165n23
　career, 157n3
　demotion of, 37
　fired by Ickes, 38, 39
　as head of USTB, 22, 24, 25, 27
　promotion of tourism by, 28
　promotion of USTB by, 23, 31
Los Angeles Times, 22
"Lost Colony" summer pageant "Negro Day," 73
Louis, Joe, 43
Louisiana Tourist Bureau, 70
Louter, David, 28
Lucander, David, 32
luggage industry, 71
"luxury" tourist trade, 157n3
Lynn, Laurence E., Jr., 2

Macnamee, W. Bruce, 173n61, 184n87, 187n21
　appointed USTB director, 38
　bureaucratic issues, 39
　career, 165n20, 185n2
　death of, 185n95
　divorce, 185n95
　praise of, 96–97, 99–100
　suggested for Board of Economic Warfare, 84–85
　Travel America and, 50
　travel expenses, 183n70
　travel promotion by, 55, 70, 74–76, 78–80
　USTB budget hearings and, 66, 68
　USTB closing recommended by, 96–100
　USTB defunding and, 98–100
　USTB headquarters office, 51
　USTB legislation and, 47, 164n9
　USTB postwar reopening and, 101
　USTB reestablishment proposed by, 101
　USTB structure and, 69, 88, 89, 95
　USTB wartime funding and, 92
　USTB wartime restructuring and, 88, 89
　wartime public relations, 82–85, 90
　on wartime travel and tourism, 50, 82–85, 87–88, 95–96
Mak, James, 6
Malek, Fred, 127, 194n57
Manchester [NH] Union, 86–87
maps, 41, 53, 70, 122–124, *following p. 62*
　Alaska, 122–123
　pictorial map of US, 53, *following p. 62*
　recreation areas, 106
　USGS, 138
Marshall, Leroy, 20
Marshall Plan, 113, 190n1
Maryland, 95–96
Mashaw, Jerry L., 2
Matsunaga, Spark, 126, 132
maximum plausibility, logic of, 13–14
McCamy, James, xii
McCarran, Pat, 68
McCarthy, Eugene, 119, 191–192n24
McDonald, Bruce D., 2
McDowell, Charles A. R., xi, 176n48
　civil rights and, 146
　Mound Bayou and, 168n59
　released from USTB employment, 73–74
　travel by African Americans promoted by, 32–33, 41–43, 56–57, 72–74

McGarry, William A., 20, 157n3
McKenzie, Aida, *following p. 62*
McLennan, Sarah Elizabeth, 3
McNabb, David E., 12
McNutt, Paul, 46, 76–80, 82
McReynolds, William, 162n40
Metropolitan Museum of Art, 163n62
Mexico, 100
Meyers, Roy T., 2
military services
    employment, 95
    non-recreational travel and, 170n7
    recreational opportunities, 142
Miller, William J., 109
Miss America pageant, 90
Mississippi Advertising Council, 168n59
Miss Sepia-America contest, 57
Moffett, Toby, 149
Monday holidays, 71, 144
Montana, x, 28
    wartime travel and, 84
    Welcome to Montana posters, 53, *following p. 62*
Montgomery, Isaiah, 168n59
Morton, Rogers, 194n63, 195n68, 195n70, 195n71, 196n75
    appointed Interior secretary, 128
    Hartzog fired by, 131
    transfer of USTB to Commerce Department by, 138
    USTB and, 128–131
Mound Bayou, 168n59
Museum Division, NPS, 24
Mussolini, Benito, 141

Nathan, Richard P., 132
National Airport, Washington DC, 76
National Archives II, 13
National Association for Motor Bus Operators, 109
National Association of Travel Officials, 105, 108, 109

National Association of Travel Organizations (NATO), 116, 121, 185n95, 187n20
National Bus Traffic Association, 68, 144
National Emergency Council, 140
National Federation of American Shipping, 109
National Luggage and Leather Goods Week, 71
National Oceanic and Atmospheric Administration (NOAA), 130
national parks
    African Americans encouraged to visit, 33, 147
    foreign visitors and, 115–116, 118
    as primary travel destination, 124
    promoting visits to, xii, 6, 11, 15, 31, 39, 79, 118, 136
    Roosevelt's expansion of, 141
    USTB posters, x, 52–53, *following p. 62*
    wartime operations reductions, 88
    wartime visits to, 50, 84, 93
National Park Service (NPS). *See also* Interior Department; United States Travel Bureau (USTB)
    Division of Tourism, 120–121
    Johnson's domestic travel promotion, 115–122
    Museum Division, 24
    Nixon and, 128–132
    postwar travel promotion by, 115–122
    purpose of, 21
    recreational facilities survey, 22
    reprogramming funding for, 118–119
    Travel Division proposal, 17–18
    Travel Office, 12
    USTB and, 6–7, 33–34, 45, 51–52
    USTB closing and, 98

USTB removal from, 37–38
wartime operations, 88, 92
Washington, D.C. tourism pilot project, 117
National Radio Forum, 30–31
National Recovery Administration, 165n20
National Resorts and Parks Association, 20
National Tourism Organization, 133
National Tourism Policy Act, 133
National Tourism Resources Review Commission, 125, 131
National Travel and Tourism Office, 134
National Weather Service (NWS), 5
Natural Resources Department, 128, 195n70
"Negro Day"
  "Lost Colony" summer pageant, 73
  New York World's Fair, 57
  Texas State Fair, 43
Negro Hotels and Guest Houses, 72
Negro Motorist Green Book, The (Green), xi, 56, 146, *following p. 62*. See also *Green Book, The*
Negro National Business, 42
Newbold, Stephanie P., 2
New Deal
  business attitudes toward, 35–37, 142–143, 152
  criticism of, xi, xii, 24, 25, 29–30, 57, 91–92, 137
  federal government expansion under, 4, 19, 29, 30, 57
  positive newspaper coverage of, 30
  proper role of government and, 148
  USTB and, 52, 140–144
New Deal coalition, 32
New England, 28
New Mexico Tourist Bureau, 70
newsletters, 40–44

newspapers
  endorsement of USTB by, 44
  praise of Macnamee by, 97, 99
  public relations via, 30–31, 123
  on wartime tourism, 86–88, 91
New York Herald Tribune, 24, 26, 31, 83, 86, 114
New York Journal-American, 86
New York State Chamber of Commerce, 157n3
New York Times, 19, 22–23, 44, 67, 70, 84, 91, 101, 161n33, 170n5
New York Tribune, 159n28
New York World's Fair, 40, 42
  "Negro Day," 57
Nicholson, Frank S., *following p. 62*
Nixon, Richard M., 119
  Action reorganization proposed by, 195n66
  Cambodia invasion under, 127
  Community Development Department proposed by, 194n61
  executive branch reorganization by, 128–132, 195n70
  Hickel and, 122–123, 127
  loyalty to, 129, 131–132
  National Park Service and, 132
  USTB and, 12, 14
  USTB defunding under, 7, 12
  USTB funding and, 122, 127
  USTB structure and, 12, 128–132
North American Travel Conference, 90
NPS. *See* National Park Service (NPS)
Nuremberg laws, 163n66
NWS. *See* National Weather Service (NWS)

Obama, Barack, 133–134, 179n93
Office for Emergency Management, 63, 79, 95, 156n7
Office of Civilian Defense, 64

Office of Defense Health and Welfare Services, 79
Office of Defense Transportation, 95
Office of Education, Interior Department, 41
Office of Government Reports (OGR), 87, 155*n*3
Office of International Travel, Commerce Department, 114
Office of Management and Budget (OMB), 130, 194*n*57
Office of Price Administration, 63
Office of Production Management, 63, 78
Office of Tourist Development, 120, 129
Office of Travel and Information Services, 14, 119–122
  advisory board, 121–123
  funding of, 120, 122
  Nixon's lack of interest in, 128–129
  nomenclature, 121
  politics and, 122
  public relations activities, 123–124
  transfer of role from Interior to Commerce Department, 132
Office of Travel Promotion, 133–134
Official Bulletin of the United States Travel Bureau, National Park Service, 13, 31–32, 40–41, 43, 54, 56, 70, 71, 75, 80, 83, 85–88, 106, 144
Ogg, Frederick A., 165*n*20
O'Malley, Thomas, 19

Pan American Airlines, 100
Parkinson's law of triviality, 174*n*7
parks. *See also* national parks
  benefits of, 11
  NPS survey of, x–xi, 22
patriotic travel, xii, 15, 30, 50, 75, 77, 79, 91, 92, 142

Pearl Harbor, 2, 7, 74, 80, 81–85, 96
Pfiffner, James P., 2
Philadelphia Tribune, 176*n*48
photographs, 69–70, 95
pictorial map of US, 53, *following p. 62*
Pillen, Cory, 3, 53
place branding, 10
political science, 138–139
politics
  Interior *vs.* Commerce department roles, 124–127
  Office of Travel and, 122
  travel and tourism promotion and, 18–19
  USTB closing and, 99–100
  USTB funding and, 9, 52, 58–59
  USTB location and, 44–47
pollution prevention, 106
Popp, Richard K., 3
Portuguese-language travelogues, 52, 89, 141
posters, ix–x, 41, 52–53, 70, 89
  See America campaign, *following p. 62*
Praeger, 3
press releases, 31, 36, 41
print materials, 123
private-sector interests. *See also* business organizations; travel industry
  advisory boards and, 133
  cooptation and, 57–58
  government contracts and, 145
  lobbying by, 9
  proper role of federal government and, 132, 149
  travel promotion and, 24, 149
  USTB and, 35, 41, 44, 128, 138–139, 142
  USTD publications and, 187*n*20
Procurement Office, 78

promotional materials
  for African Americans, 146
  audio-visual activities, 69
  brochures, 40, 41, 51
  Calendar of Events, 40, 70, 88
  color slides, 69
  film loans, 69
  free and low-cost, 123–124
  government- *vs.* industry-produced, 152
  lending library, 54
  maps, 41, 53, 70, 106, 122–124, *following p.* 62
  for military personnel, 95
  movies, 41
  newsletters, 40–41, 42–43, 44
  Official Bulletin, 13, 31–32, 40–41, 43, 54, 56, 70, 71, 75, 80, 83, 85–88, 106, 144
  photographs, 69–70, 95
  pictorial map of US, 53
  Portuguese language materials, 52, 89
  posters, ix–x, 41, 52–53, 70, 89, *following p.* 62
  press releases, 31, 36, 41, 85
  publications, 40–41
  See America posters, x
  Spanish language materials, 52, 89, 141
  state guidebooks, 52, 54
  Travel and Recreation News Letter, 40–41, 55–57, 71
  traveling exhibits, 70
  Travel News, 40, 56, 71
  travelogues, 69, 89, 95
  Travel USA Bulletin, 106, 187*n*20
  Travel West, 71, 81, 84, 88
propaganda, 5, 25, 58, 104, 137
public administration
  executive branch reorganization, 128–132
  historical research on, 2–3
  interest in, 165*n*20
  literature review, 2–3
  public relations in, xii
  USTB as study in, 136–139
Public Administration Review, 12
public interest
  industry advisory boards and, 151
  private interests *vs.*, 44
  travel promotion and, 148–151
public policy, 9–10
public relations (PR)
  Congressional opposition to, xii, 5–6, 58
  controversial use of, 10–11
  free publicity, 123
  newspapers and, 30–31, 70, 84–85
  Office of Travel, 123–124
  after Pearl Harbor, 81–85
  as propaganda, 5, 25, 58, 104, 137
  in public administration, xii
  radio and, 29, 30, 36, 54, 66, 70
  radio use, 29, 30, 36, 54, 66, 70, 123
  "See America" campaign and, 53, 116
  "See America First" campaign, x, 24, 40
  "See the Old West This Year" campaign, 71
  "See what you defend" motto, 80
  speeches, 70, 90–91, 106
  third-party information dissemination, 5
  "Travel Strengthens America" campaign, 74–75
  "Travel with a Patriotic Purpose" campaign, 75
  USTB activities, x, 1, 5–6, 29–32, 40–41, 52–56, 69–72
  USTB legacy and, 152
  USTD activities, 106

public relations (PR) *(continued)*
　in wartime, 82–85, 90–91
　Western states tourism promotion, 28
public service announcements, 123
Public Works Administration (PWA), 30, 46, 162*n*40
Puerto Rico, 23, 28–29, 35, 92
PWA. *See* Public Works Administration (WPA)

Question Box Program, 30

Raadschelders, Jos C. N., 2
racism, x, xi, 11, 32, 33, 43, 57, 59, 72, 144–147, 176*n*48. *See also* African Americans; civil rights
radio promotion, 29, 30, 36, 54, 66, 70, 123
railroads, 1, 50, 88, 90–91
Railway Age, 95
Rayburn, Sam, 18
Reagan, Ronald, 119, 133, 149
recreational facilities survey, 22
Recreation Areas of the United States under Federal or State Administration, 106
Reed, Nathaniel, 194*n*58, 194*n*63, 195–196*n*71
Rees, Edward, 93, 94, 183*n*61
relief programs, ix, 4, 16, 17, 22, 39, 47, 51, 61, 77
reprogramming, 118–119
Republican National Committee (RNC), 128, 165*n*23
Republican party
　African Americans and, 32
　attitudes toward USTB by, 19–20, 107–110
　business interests and, 103–104
　early proposals for travel and tourism office, 15–17

　federal government expansion and, 19–20
　government spending and, 103
　on proper role of federal government, 149
　USTB funding and, 91–92
research methodology and sources, 12–14
research reports, 55
Revella Hotels, 176*n*48
Rinella, Michael, xiii–xiv
Robinson, Joseph, 19
Rockefeller, Nelson, 119
Rockwell, Stephen J., 3
Rohde, Ruth Bryan, 37, 38–39, 165*n*14, 166*n*31
Rommel, Erwin, 63
Romney, George, 192*n*25
Romney, Mitt, 192*n*25
Room to Roam, 123
Roosevelt, Eleanor
　civil rights and, 144–145
　on domestic tourism, 36
　Rohde and, 37, 38, 39, 165*n*14
Roosevelt, Franklin D., 172*n*47
　on 1939 world's fairs, 40
　agency name system, 159*n*6
　appropriations and, 155*n*3
　business attitudes toward, 44, 152
　on civil rights, x, 32, 144–145
　conservation policies, xii, 50
　cooptation and, 143–144
　death of, 102
　early initiatives, 17
　Economic Defense Board, 179*n*7
　elected for second term, 20
　as enemy of free enterprise, xi
　federal government expansion by, 4, 19–20, 29, 30, 57–58, 142–143
　Garner and, 186*n*4
　Gerard's resignation letter to, 28

Good Neighbor policy, 24, 29, 35, 49–50
  Greeley joke, 26, 159n28
  Ickes and, 77, 178n74
  Latin American relations, 52
  legacy of, ix, 140–144, 151
  limited national emergency declared by, 50, 64, 142, 173n2
  McNutt and, 76–80
  New York Tribune and, 159n8
  opposition to, 97, 103, 148
  public administration and, 2
  public lands and, 11–12
  public relations programs, 1, 10–11, 58
  public welfare department plans, 178n69
  third term, 2, 10, 50, 61, 186n4
  unlimited national emergency declared by, 64, 75, 80
  USTB creation and, 1–2, 10, 20, 61
  USTB funding and, 68
  USTB legislation and, 59
  USTB policy goals and, 140–144
  USTB promotion and, 25–26, 29, 137–138
  wartime budget, 63–64, 65, 91
  on wartime travel, 49–50, 79, 88, 181n23
  WPA grants and, 39–40
Roper, Daniel C., 18, 19, 20
Rosenbloom, David H., 2
Rothstein, Jerome (aka Roth), *following p. 62*
rubber shortages, 78, 83
Rubin, Irene S., 2
Ruedi, Oreen M., 70

safety, of wartime travel, 76
San Francisco world's fair, 28, 36, 40, 55

Santa Fe Railroad, 70, 84, 90–91
Santoro, Vincenzina, 115
Saturday Review, 121
savings bonds, 83
Sawyer, Charles, 109
Saylor, John, 192n32
Schachter, Hindy Lauer, 2
Schmeckebier, Laurence F., 3
Schultz, George, 195n70
Second Inter-American Travel Congress, 68
Secretary's Order 2912, 120–121
"See America" campaign, x, 53, 116, *following p. 62*
"See America First" campaign, x, 24, 40
"See the Old West This Year" campaign, 71
"See the United States" campaign, 116
"See what you defend" motto, 80
segregation, 42, 43, 72, 146–147. *See also* African Americans
  in hotels, 176n48
Sellers, Richard West, 6
Selznick, Philip, 143
Semmens, Kristin, 3, 10
Senate Appropriations Committee, 64–65, 68, 93, 102, 110–111, 118
Senate Commerce Committee, 17, 19, 34, 45, 47–48, 58, 59–60, 126
Senate Interior Committee, 9, 192n32
Senate Travel Caucus, 133
"separate but equal" principle, 146–147
Seventy-Fifth Congress, 21–22, 33–34, 43
Seventy-First Congress, 15–16
Seventy-Fourth Congress, 17–20
Seventy-Ninth Congress, 102–103
Seventy-Second Congress, 16–17
Seventy-Seventh Congress, 64–68
Seventy-Sixth Congress, 1–2, 4, 43–48, 57–61, 169–170n81
Seventy-Third Congress, 17

Shaffer, Marguerite S., 1, 3
Shooshan, Harry M., 117, 191*n*21
skiing, 54, 55, 70
Sklaroff, Lauren Rebecca, 32
Smith, Cyrus, 122
Smith, Harold, 172*n*47, 184*n*80
Smith, Luther Ely, 89
Sorin, Gretchen Sullivan, 3
Sources of Travel Information, 106
South America, 52, 69
Southern Pacific Railroad, 90–91
southern states
   civil rights and, x, 144–145
   places of interest to African
      Americans, 56
   promotion of, 28
Soviet Union, 48, 63, 186*n*4
Spanish-language travelogues, 52, 89, 141
spas, 74
Staggers, Harley O., Jr., 117, 125
Stalin, Joseph, 169*n*77
Stans, Maurice, 123, 129, 138, 193*n*41
State Department
   budget review by, 68
   foreign tourism policy, 156*n*9, 163*n*66, 168*n*70, 173*n*55
   support testimony by, 44
   USTB and, 7
state guidebooks, 52, 54
state tourism/travel offices, 29, 41
State Travel Information Agencies (USTB), 40–41
Statue of Liberty, 89
Stivers, Camilla, 2, 3
student dissent, 127
summer trips, 88
Swain, Donald C., 3, 7
Swanson, Jason R., 6, 7, 133

Taylor, James, 20

television, 123
Tennessee Valley Authority, 143
Terry, Larry D., 2
Texas State Fair "Negro Day," 43
This Week, 106
Thomas, Lowell, 30, 163*n*50
Thornton, Robert L., 115
three-day weekends, 71, 144
tire rationing, 78, 85, 90
Tolson, Hillary A., 101, 184*n*88
tourism. *See also* domestic tourism;
    foreign tourism; travel
   benefits of, 20, 28, 30, 75, 76, 79–86, 90–91
   business interests and, 15–17
   Congressional interest in, 15–20
   conservation and, 141
   dude ranches, 90
   economic benefits of, 82–83, 116, 140, 143, 148–151
   economic interests and, 15–17, 18–19, 30, 75
   employment by, 150
   in Europe, 113–114
   federal government and, 1–2, 15–26, 147–151
   foreign visits to US, 67, 115–117
   gasoline supply and, 77–78
   by German Jews, 156*n*9, 163*n*66, 168*n*70, 173*n*55
   inflation and, 83
   inter-American, 35–36, 52
   Latin America and, 141
   literature on, 3
   "luxury" tourist trade, 157*n*3
   as patriotic, xiii, 15, 30, 50, 75, 77, 79, 91, 92
   research, 55
   rubber shortages and, 78, 85
   safety issues, 76
   Southern states, 28
   summer trips, 88

Index 233

trade deficit and, 116–117
wartime, 50, 74–80, 79
wartime USTB programs, 81–85
Western states, 28
World War II and, 35–36
year-round, 106
Tourism Policy Council, 133
Tourism Review Commission, 128
trade deficit, 116–117
Traffic World, 19
Transportation Department (DOT), 117, 191n19
travel. *See also* tourism
   benefits of, 50, 79–80, 81–85
   gasoline supply and, 77–78
   government promotion of, 4–6
   Great Depression and, xii
   inflation and, 83
   national health and, 77
   politics and, 19
   safety concerns, 76
   USTB's wartime programs, 81–85
   wartime economy and, 83
   wartime effects on, 74–80, 86
travel agencies, 67, 104, 106, 177n64
Travel America, 48, 49–50
Travel America Hall, 55
Travel and Recreation News Letter, 40–41, 55–57, 71
"Travel Filler," 123
travel films, 69
travel guidebooks, 54
travel industry. *See also* business organizations; private-sector interests
   advisory boards, 104–105, 120–123, 151
   consumer spending and, 83
   cooptation with federal government, 143–144, 151
   economic recovery and, 140–144
   employment by, 150
   for-profit, 41
   government promotion of tourism and, 114–115
   lobbying by, 38, 143–144
   materials distributed by, 66
   one-on-one outreach to, 84
   promotional materials produced by, 152
   proper role of federal government and, 147–151
   public interest and, 148–151
   Republican party and, 103–104
   self-interests of, 80, 90–91
   speeches to, 55
   support for government travel promotion by, 113, 133
   support for USTB by, xi, 57, 67–68, 142–144
   support for USTD by, 107–108
   USTB closing and, 97–98
   USTB coordination, 57–58, 66
   USTS and, 187n20
   during wartime, 82–83, 90–91
Travel News, 40, 56, 71
"Travel Notes" (USTB), xi
Travel Office, NPS, 12
travelogues, 52, 69, 89, 95, 141
travel posters, 52–53
Travel Promotion Act, 133–134, 196n85
travel reporters, 105
travel research, 55, 106
travel reservations, 76, 177n64
Travel Review Commission, 129
"Travel Strengthens America" campaign, 74–75
Travel USA Bulletin, 106, 187n20
Travel West, 71, 81–82, 84, 88
"Travel with a Patriotic Purpose" campaign, 75
travel writers, 75
triangulation, 12, 13

trucking industry, 36–37, 87
Truman, Harry S.
 election of, 110
 Marshall Plan and, 190*n*1
 Republican opposition to government programs maintained by, 103
 Roosevelt's selection of for vice president, 186*n*4
 USTB and, 10, 14, 102–103, 107–108
 Wallace and, 186*n*4
Trump, Donald, 134

Udall, Stewart
 Hartzog and, 131
 Johnson's decision not to run and, 118
 political power struggles and, 127
 travel task force, 116–117
 USTB funding and, 119–122, 191*n*21
 USTB revival and, 13, 192*n*31
unemployment, 18, 50. *See also* employment
United Kingdom, 82
United Press, 38, 89, 90
United States Chamber of Commerce, 15, 44, 103, 143
United States Tourist Bureau (USTB), xvi, 6, 14
 change of name to United States Travel Bureau (USTB), 27–28
 establishment of by secretarial order, 1, 13
United States Tourist Information Bureau, 22
United States Travel Bureau (USTB), 56, 89, 109. *See also* National Park Service (NPS); promotional materials
 African American travel promoted by, x–xi, 32–33, 41–43, 144–147
 appropriation of funds for, 64–68
 archival sources, 12–13
 authorization of, 1–2, 4, 57–61, 64
 budget, 39–40, 51, 65–66, 69, 74, 88–90, 96, 98–99, 150
 business support of, xi–xii, 4, 35–37, 104
 chronology of, 12, 14
 closing of, 96–100, 101, 135, 182*n*53
 collaborators, 23–24, 28, 32, 37, 72
 Commerce Department and, 7, 33–34, 44–47, 102, 113
 Congressional disputes over, 91–95
 defunding of, 7, 93
 Division of Negro Affairs, xi, 42, 72, 146
 environmental protection and, 11–12
 establishment of, 1–2, 4, 21–23, 26
 federal government expansion and, 4, 19–20, 29, 30, 57
 federal government role and, 147–151
 field offices, 28, 65–66, 89, 91, 182*n*41
 funding process, 2, 9, 102–104, 135
 historical overview, 135
 historical significance, 139–151
 as information clearinghouse, 16, 24, 54
 Information Section, 24
 jurisdiction, 22–23, 44–47
 Latin America and, 6, 7
 law creating (1940), 1–2, 6–8, 14, 51–52, 57–61, 64, 115–126, 129, 131, 132, 135, 136, 138
 leadership of, 27–29, 37–40, 51–52, 68–69, 88–90
 legacy of, 3–4, 8–12, 135–153
 legal status termination, 14

legislative support of, 33–34
mission of, xii, 1–2, 5, 51, 52, 136–137, 148, 152
New Deal policy goals and, 140–144
New York office, x–xi, 23, 27–28, 32–33, 40, 41–43, 51, 55, 56–57, 69, 70, 71, 72–74, 78, 88–89, 90, 97–98, 146, 161*n*33, 175*n*33, 182*n*41
nomenclature, 6, 14, 27–28, 105, 121, 136
NPS and, 6–7
Operations Section, 24
organization of, 27–29, 37–40, 51–52, 68–69, 88–90, 103–104, 105, 136
origins of, 15–17
overview, 1–14
political science and, 138–139
politics and, 9, 44–47, 52, 58–59, 99–100
postwar revival of, 101–111, 102–104
as a propaganda agency, 137
as public administration study, 136–139
public interest and, 148
Publicity Section, 24
as public policy study, 9–10
public relations role, x, 1, 5–6, 10–11, 29–32, 40–41, 52–56, 69–72, 137, 152
published errors about, 6–8
Republican party and, 15–17, 103–104
research methodology and sources, 12–14
research reports, 55
Roosevelt's praise of, 29
San Francisco office, 28, 37, 40, 51, 54, 55, 56, 69, 70, 71, 74–75, 79, 84, 88–89, 146, 182*n*41

special interests and, 138–139
staff, 39–40, 51, 65–66, 68, 88–89, 92, 93, 95, 136, 175*n*33
State Department and, 7
transferred back to NPS, 51–52, 57
transferred to Office of the Secretary, 37–38, 51
travel industry and, 66, 67–68, 80
wartime budgets, 91–95
wartime operations, 74–80, 81–100
wartime shutdown of, 7
Washington headquarters office, 27, 38, 39, 51, 88
United States Travel Division (USTD), xvi, 6, 14
advisory committee, 104–105, 106, 108
appropriations process, 107–111
business support for, 107–108
closing of, 110–111, 115, 189*n*45, 193*n*42
collaborators, 105, 158*n*12
conservation policy goal, 12, 106
Democratic opposition to, 110–111
environmental movement and, 106
establishment of, 104–107
funding of, 107–110, 110–111
industry support for, 107–108
law authorizing, 111
name change, 105
offices, 105–106
public relations activities, 106
Republican support for, 107–110
staff, 105–106
travel industry interests and, 187*n*20
year-round tourism encouraged by, 106
United States Travel Service (USTS), 115, 120, 123
Carter's defunding of, 133
funding, 124–126

United States Travel Service (USTS) *(continued)*
  Interior Department's jurisdiction over, 187*n*13
  mentioned by AP, 188–189*n*39
  Morton's oversight of, 195*n*70
  Reagan's elimination of, 133
Universal Service, 165*n*20
unlimited national emergency declaration, 64, 75, 80
Urban League, 33
US Civil Service Commission, 69
US Commission on Organization, 110
US Congress (source), 70
US Department of Agriculture, 4
US Domestic Travel Act, 7
US General Accounting Office (GAO), 98
US Geological Survey, 3, 122–123, 138, 159*n*25
U.S. Government Reports, 70
US House (source), 15, 16, 19, 20, 23, 38, 39, 40, 44, 45, 47, 51, 52, 54, 61, 65, 66, 67, 68, 69, 90, 91, 93, 94, 95, 98, 103, 104, 105, 108, 109, 110, 114, 117, 118, 122, 123, 124, 125, 126, 129, 130, 131, 132, 133, 148, 149, 150
US Maritime Commission, 36, 44, 78, 170*n*5
US National Tourism Resources Review Commission, 131
US NPS (source), 20
US OMB (source), 128
US Postal Service commemorative stamps, ix–x, 53, *following p.* 62
US Senate (source), 19, 34, 45, 47, 52, 60, 68, 93, 94, 102, 103, 106, 114, 118, 124, 125, 127, 129, 144

USS *Greer,* 64
USTB. *See* United States Travel Bureau (USTB)
USTB (source), xi, 29, 42, 51, 53, 55, 70, 72, 88, 106
USTD. *See* United States Travel Division (USTD)
USTD (source), 105–107
US Travel and Tourism Administration, 133
US Travel Commission proposal, 17–18
US Virgin Islands, 29

Valin, Jean, 10
Vandenberg, Arthur, 60
van Thiel, Sandra, 12
V-days, 79
Velonis, Anthony, 53
Vietnam War, 127
Virginia Chamber of Commerce, 70
Visitours, Inc., 157*n*3
Visit the National Parks poster, *following p.* 62

Wadsworth, James, Jr., 19
Walker, Ronald H., 132, 196*n*74, 196*n*75
Wall, Robert, 188*n*46
Wallace, Henry, 83, 102, 186*n*4
wartime travel and tourism. *See also* World War II
  benefits of, 82–83, 85–86, 90–91
  consumer spending and, 83
  criticism of, 86–88
  European tourists and, 35, 49
  gasoline supplies and, 77–78, 85, 86, 178*n*69, 181*n*23
  in Germany, 82
  Ickes's contradictory views on, 85–88
  inter-American, 35–36

Macnamee's views on, 83–85, 95–96
public relations and, 83–85, 90–91
rubber shortages and, 78, 83
speeches promoting, 90
travel industry and, 48, 82–83
travel industry self-interest and, 90–91
travel promotion and, 81–100
in UK, 82
USTB and, 81–85
Washington Post, 31, 44, 96, 111, 114, 156*n*7
Washington Star, 30–31
weather bureau, 5
Weber, Gustavus A., 3
weekend travel
  by military, 170*n*7
  Monday holidays and, 71, 144
  by truckers, 36, 87
Weitzman, Martin, *following p.* 62
Welcome to Montana poster, x, 53, *following p.* 62
western states, 28, 71
West Virginia Conservation Commission, 22
Wheeler, Burton, 169*n*76
White, Richard D., 2
Williams, Daniel W., 2
Williams, Wythe, 31
Wilson, James Q., 137, 143
Wilson, Woodrow, 37
Wingate, Jay, 69, 88, 89, 182*n*53
Winston-Salem [NC] Journal, 44
Wirth, Conrad, 27, 161*n*26
Wisconsin Historical Society, 175*n*30
Witham, Glenn, 7
Wolverton, Charles, 109
Women's Wear Daily, 53, 55
WOR, 24

Works Progress Administration (WPA), 23, 46, 47, 51, 69, 135, 162*n*40, 175*n*33
  commemorative stamps, ix–x, 52–53
  Federal Writers' Project, 22, 40, 54
  graphic artists, ix–x
  posters, 52–53, *following p.* 62
world's fairs, 28, 36, 54
World War II. *See also* wartime travel and tourism
  domestic policy changes and, 82
  European tourists and, 35, 49, 141–142
  European travel and, 141–142
  gasoline supplies, 77–78
  inter-American travel and, 35–36
  legislation related to, 61
  rubber shortages, 78, 86
  tourism and, 35–36
  travel and tourism during, 35–36, 48, 74–80
  travel promotion during, 81–100
  US response to, 63–64
  USTB during, 74–80, 81–100
  USTB public relations during, 81–85
  USTB shutdown during, 7
WPA. *See* Works Progress Administration (WPA)
writers, 22, 40, 54, 75
Wrobel, David M., 3

Yellowstone Park, 116
YMCA/YWCA, 42, 72
Yosemite National Park, 79
Your Government at Your Service, 30

Zavattaro, Staci M., 10, 134
Zelizer, Julian E., 2, 10, 17

www.ingramcontent.com/pod-product-compliance
Lightning Source LLC
Chambersburg PA
CBHW030536230426
43665CB00010B/913